"This fascinating study of Chinese wind and solar technologies uses an STS approach to study the Chinese governance of energy transition. It gives a highly original perspective on China's transformation from a manufacturing-based economy into a design and innovation country, and raises important questions about the possible directions of China's future development."

Wiebe E. Bijker, Professor Emeritus of Technology and Society, Maastricht University, The Netherlands

"A renewable energy revolution is happening in China, and this book provides an excellent backdrop to understand how this revolution has come about. Given the importance of understanding China's role in fighting the global climate break-down, Marius Korsnes' book is particularly timely and welcome. *Wind and Solar Energy Transition in China* provides a detailed and well-written account of the conditions within which China's wind and solar industries have grown, helping us understand how sustainable change is strategically governed in China."

Jørgen Randers, Professor of Climate Strategy, Norwegian Business School, Norway

"This book nicely summarizes the rise of China's wind and solar industries, as well as the extensive body of literature that has studied these industries over the past two decades. The examination of emerging renewable energy applications in China, including offshore wind, distributed solar and smart cities, is particularly timely."

Joanna Lewis, Associate Professor and Director of the Science, Technology and International Affairs Program, Georgetown University, USA

"This book offers one of the first detailed accounts of China's distinctive mode of governance of its high technology renewable energy industries. It shows how the Chinese government is able to orchestrate the development of wind and solar PV technology, despite the high costs and long timeframes involved. One element in this is the deployment of future expectations that mobilise investments and sustain the development of industrial capability. In contrast to accounts which emphasise centralisation of Chinese government, Korsnes' analysis offers novel insights into the productive role of regional and sectoral fragmentation and temporal delays in these processes."

Robin Williams, Professor in Science Technology and Innovation Studies, School of Social and Political Science, University of Edinburgh, UK

"In an informative and engaging way, *Wind and Solar Energy Transition in China* shows the visionary policy goals and various hurdles the Chinese government and industry actors have faced in the greening process of the world's largest electric power system. Marius Korsnes has written a carefully researched book, presenting a fascinating view to what happens behind the scenes of China's low-carbon energy transition."

Peter Karnøe, Professor, DIST – Center for Design, Innovation, and Sustainable Transition, Department of Planning, Aalborg University, Denmark

Wind and Solar Energy Transition in China

This book explores the mobilisation of China's wind and solar industries and examines the implications of this development to energy generation and distribution, innovation and governance.

Unlike other publications that focus mainly on the formal policy landscape and statistics of industry development, this book delves deeper into the ways in which the wind and solar industries have evolved through negotiations made by the involved stakeholders, and how these industries play into larger Chinese development and policymaking interests. Overall, it sheds new light on the strategic development of China's renewable energy industry, the flexible governance methods employed and the internal struggles which Chinese local, regional and central policymakers, and state-owned and private enterprises have faced.

This book will be of great relevance to students and scholars of renewable energy technologies, energy policy and sustainability transitions, as well as policymakers with a specific interest in China.

Marius Korsnes is a postdoctoral researcher connected to the Research Centre on Zero Emission Neighbourhoods in Smart Cities (ZEN) and the Norwegian Centre for Energy Transition Strategies (NTRANS) at the Norwegian University of Science and Technology, Norway.

Routledge Explorations in Energy Studies

Energy Poverty and Vulnerability
A Global Perspective
Edited by Neil Simcock, Harriet Thomson, Saska Petrova and Stefan Bouzarovski

The Politics of Energy Security
Critical Security Studies, New Materialism and Governmentality
Johannes Kester

Renewable Energy for the Arctic
New Perspectives
Edited by Gisele M. Arruda

Decarbonising Electricity Made Simple
Andrew F Crossland

Wind and Solar Energy Transition in China
Marius Korsnes

Wind and Solar Energy Transition in China

Marius Korsnes

Routledge
Taylor & Francis Group
LONDON AND NEW YORK

earthscan
from Routledge

First published 2020
by Routledge
2 Park Square, Milton Park, Abingdon, Oxon OX14 4RN

and by Routledge
52 Vanderbilt Avenue, New York, NY 10017

Routledge is an imprint of the Taylor & Francis Group, an informa business

British Library Cataloguing-in-Publication Data
A catalogue record for this book is available from the British Library

Library of Congress Cataloging-in-Publication Data
A catalog record has been requested for this book

ISBN: 978-0-367-19418-5 (hbk)
ISBN: 978-0-367-19419-2 (ebk)

Typeset in Bembo
by Wearset Ltd, Boldon, Tyne and Wear

Contents

Illustrations

Figures

Table

Preface

This book is an outcome of more than eight years of research on China's wind and solar industries. I wrote my master's thesis about China's wind industry (entitled 'The Growth of a Green Industry: Wind Turbines and Innovation in China'), under the supervision of Hal Wilhite and Jens Hanson (University of Oslo). I wrote my PhD thesis about the offshore wind industry in China (entitled 'Chinese Renewable Struggles: Innovation, the Arts of the State and Offshore Wind Technology'), supervised by Knut H. Sørensen, Marianne Ryghaug and Gard H. Hansen (Norwegian University of Science and Technology). And since 2016, I have been working on my postdoc project focusing on China's development of solar photovoltaic (PV) panels, and the role of energy prosumers in China. When I started out the work on my master's thesis in early 2011 my idea was to cover both the solar and wind industries in China, and I was initially seeking interviewees from both fields. However, I ended up getting informants mainly from the wind industry, and I therefore decided to focus only on wind. That focus on wind is also how I stumbled upon the opportunity to do a PhD thesis about China's offshore wind industry. Now, more than eight years later, I have actually done what I intended to do at the outset: write a comparative study focusing on the extremely rapid growth that has happened in China's wind and solar industries. At that time, in 2011, China had already installed a lot of wind turbines, and it was far from obvious that China would start installing solar panels as well—which eventually grew even more rapidly than the wind industry had done. My idea then, was to try to understand how China had been able to become a world-leading *manufacturer* of both wind turbine and solar PV technology, but having a domestic market only for wind turbines. With a broad understanding of innovation, this book delves into the various reasons for this wind and solar energy transition happening in China.

It should be said, though, that I have also had a steep learning curve in this journey. I started out with a rather fixed understanding of what innovation entails; an innovation system understanding. When I started out trying to understand how innovation came about in China, I experienced an increasing sense that the innovation systems approach did not fit well with the Chinese experience. I did notice that there were large investments in research and development, and several companies were entering the wind industry.

Chinese companies were trying to acquire technology from European companies and used all the means they could to learn. But did this mean that they were innovating? Moreover, in the eyes of European companies, Chinese companies were only copying them, not interested in quality, taking too many risks and so on. There were several sides to the same story. Which side was right? As I got more familiar with the strategies of Chinese and European companies, I found out that they all had different goals and diverging approaches. European companies were, for instance, afraid that they would one day be out of business, as Chinese companies continued to learn. As I also became more acquainted with the science and technology studies (STS) literature, I felt that it could provide a richer set of nodes to base my analysis on. It allowed me to include a larger set of actors and motivations. At the outset of my PhD work I was also rather excited about the 'sustainability transitions' literature, and I was curious to see how well this literature fit with what I had experienced and learned in China. Ultimately, therefore, all three of these approaches have informed and shaped the resulting analysis of this book. Naturally, innovation in a Chinese context cannot be accurately described by one framework, thesis or book alone. A collective academic effort is necessary to better understand how innovation may be described in a Chinese context. I would nevertheless say that the journey I have gone through has been enriching and has given me several insights which I surely would not have acquired otherwise. Hence, the resulting analysis is presented in this book, and I hope that it will be of use and inspire a new and different way of thinking about innovation in a Chinese context.

Several parts of this book rely on work done for my PhD thesis, written at the Department of Interdisciplinary Studies of Culture, Norwegian University of Science and Technology (NTNU). Some material and ideas have been reused, but they have all been adapted and adjusted to this format and context. Some of the material is also based on my master's thesis which I wrote at the Centre for Development and the Environment, University of Oslo (UiO). Since wind been my main focus of research the past years, a majority of examples in the book relate to the wind industry. However, I have done my best to balance the focus on wind and solar. Moreover, some parts of the book build on material which I have published elsewhere. This material is

- Chapter 4: parts of the text are based on the following article: Korsnes, M., 2014. Fragmentation, centralisation and policy learning: An example from China's wind industry. *Journal of Current Chinese Affairs*, 43(3), pp. 175–205. https://doi.org/10.1177/186810261404300308. Article published open access, and copyright is still with me, the author.
- Chapter 5: parts of the text is based on a previously published article: Korsnes, M. and Ryghaug, M., 2017. With license to build: Chinese offshore wind firms rejecting European certificates. *Technology Analysis & Strategic Management*, 29(7), pp. 750–761. https://doi.org/10.1080/09537325.2016.1236188. www.tandfonline.com/. Material is reused with permission from Taylor and Francis.

- Chapter 6: parts of the text is based on a previously published article: Korsnes, M., 2016. Ambition and ambiguity: Expectations and imaginaries developing offshore wind in China. *Technological Forecasting and Social Change*, 107, pp. 50–58. https://doi.org/10.1016/j.techfore.2016.03.030. Material is reused with permission from Elsevier.

I am extremely grateful to the support, supervision and encouragement I have gotten through writing both my master's thesis and PhD thesis. The ideas and analysis presented in this book would not be half as good without the guidance, patient reading and commenting of my brilliant supervisors Hal, Jens, Marianne, Knut and Gard. I also spent in total almost two years in China as part of fieldwork for this research. In addition to many shorter stays, I spent 5 months at the Beijing Language and Culture University in 2011 11 months at the Shanghai Jiaotong University in 2013 and 2014, and 3 months at Tsinghua University in 2018. I am greatly indebted to all the friends and colleagues I met and that helped me out so much during these stays. Without your support, encouragement, hospitality and friendly attitudes, none of this work would have been possible at all. The open and warmhearted way I have been received anywhere I travel in China is indeed one of the primary reasons why I cannot wait to come back again and again.

At the Shanghai Jiaotong University I would like to thank the people at the Department for Mechanical Engineering, particularly Professor Ruzhu Wang, Professor Yong Li and Professor Hu Xiajun for great hospitality, and collaboration through the Sino-Norwegian Partnership on Sustainable Energy (SiNoPSE). At Tsinghua University, I would like to thank my colleagues at the Institute of Energy, Environment and Economy, particularly Professor Teng Fei, Dr. Wang Yu and Su Xin, also partners in SiNoPSE. I would also like to thank Professor Hong Wei and Professor He Jijiang at the Institute of Science, Technology and Society, Tsinghua University, as well as Professor Liu Jian at the School of Architecture, Tsinghua University. I am also indebted to several people who have helped, commented and discussed this topic with me throughout the years. I have been lucky to know and discuss with Jorrit Gosens. Jorrit was also one of the initiators of the extremely useful webpage chinaenergyportal.org, which translates central, energy-related Chinese policy documents and statistics into English. The GIZ has also been very helpful, in particular Anders Hove, and the Energy Transition China newsletter. I am also grateful to Julia K. Kirkegaard who has written an excellent book on China's wind experiences, and whom I have learned a great deal from. More recently, I have also had the great pleasure of getting to know David Tyfield, whose insights and perspectives on innovation in China are quite intriguing. Iselin Stensdal and Anna Korppoo have also been great partners for discussions on China, institutions, climate and energy issues. I am grateful for the great community at the Norwegian University of Science and Technology (NTNU) that really caters for great collaboration with Chinese partners. In particular, Wang Yu, Annemie Wyckmans, Pål Aune and Arne Bredesen have been helpful and indispensable to my engagement with China. I owe a big thanks to Sophia Efstathiou and Laura Sommer, and several other people who occasionally joined the "shut-up and write" sessions which helped me write this book. Also, all colleagues at the Department of Interdisciplinary Studies of Culture have been exceptionally supportive and stimulating—thank you for creating a great work environment! In particular, I would like to thank my colleagues Thea Marie Valler and William Throndsen for providing excellent feedback and comments for several chapters of this book. I am also grateful to my wonderful friends and family for their support and encouragement. Lastly, I would like to thank Sílvia Coutinho for all her love, support and encouragement, for great discussions about China and the world, and very useful feedback to this book.

I am always happy to get feedback and I would love to hear potential opinions about what I have written in this book. Don't hesitate to write me an email! In this way we can continue the discussion and continue to improve our understanding of the wind and solar energy transition in China. I hope you enjoy reading the book!

1 Introduction

The transformation which is taking place in the People's Republic of China (hereafter China) is unrivalled in history. China's economy has grown faster for a greater length of time than has any other country (Lardy 2019), and this has had a major impact on China's population, which amounts to almost one-fifth of the global population. The opening-up and reform which started under Deng Xiaoping in 1978 has affected every aspect of life in China, from individuals, families, societies, governments and international relations to pollution, greenhouse gas emissions (GHG) and the destruction of natural resources (Saich 2011; Shapiro 2012). These changes have brought with them both opportunities and threats for China and for the world. China is becoming more interested in the rest of the world and vice versa. But how do experiences, knowledge and technology amalgamate into sustainable industries?

This book examines how the wind turbine and solar photovoltaic (PV) industries have developed in China.[1] Comparing and contrasting these two renewable energy industries, the book sheds light on innovation processes more generally in the context of China, which is important in order to say something about whether these industries ultimately represent a transition to a more sustainable society. Innovation is here viewed broadly, incorporating interactive learning and circulation of knowledge, competences, experiences and expectations connected with wind and solar technology. A wide definition of innovation allows for an analysis of wind and solar technology which is not only limited to the technology itself, but also amongst other things includes social, economic, environmental and political aspects. This book therefore shows how innovation and learning are politics-imbued processes.[2] The innovation process does not happen in a vacuum, but it is embedded in complex negotiations of power, pride and culture that transcend the technology in question.

In May 2019, China launched the first ever batch of nonsubsidised wind and solar PV power generation projects (so-called grid-parity projects) (chinaenergyportal.org 2019a). This included almost 21 gigawatt GW of wind and solar PV projects approved by the central government, spread out over 16 provinces and operational by 2021. All of them are providing electricity to prices comparable to coal power. In order to achieve this regional governments, government banks and grid operators were mandated to help out by providing long-term (20-year) contracts and minimising nontechnical costs (power-technology.com 2019). What is more, the government also announced that from 1 January 2021, onwards, all onshore wind projects would reach 'grid-parity' and no longer receive any subsidies. Only a few years ago, it would have been unthinkable that renewable energy would be able to compete with coal energy in China or even globally. Since 2013, China has taken several measures to reduce fossil fuel subsides (Yuan *et al.* 2019). One of the most important changes is related to a planned deregulation of China's electricity system in 2020, entailing a change away from the set benchmark price of electricity and the planned annual operating hours policy—which historically has given considerable benefits to coal power production (Yuan *et al.* 2019).

China did go through a round of electricity sector deregulation in the early 2000s, when electricity generation companies were separated from the electricity transmission and distribution grid. Since then, as China's economic and electricity consumption growth has slowed, talk of deregulation has been renewed, given the danger of producing too many coal power plants ending up with a highly inefficient electricity system and a risk of stranded assets—given that the lifetime of an average coal-fired power station is 40 years (Wei *et al.* 2018). With the so-called Document No. 9, which was launched in 2015, the deregulation process was expedited, with intentions to gradually introduce market-based principles for electricity pricing from 2020. This change will likely be beneficial for China's solar PV and wind industries, as it implies a removal of several indirect coal power subsidies (Yuan *et al.* 2019). As pointed out by Li Junfeng, a government official very central to China's renewable energy development, at the 2018 China Offshore Wind Summit: 'President Xi Jinping has recently said that coal power must be reduced. This means that in the future, wind will not compete with coal, but with other renewables'.

You might wonder: Why focus on wind and solar energy, when coal power still is so much bigger in China? It is true, a low-carbon energy transition in China implies a reduction in the dominance of fossil fuels at the same time as renewable energy technologies grow.[3] In fact, it is estimated that around 40 percent of all domestic Chinese emissions and 11 percent of global CO_2 emissions come from Chinese coal power plants (Alva and Li 2018). For this reason, it is important to focus on ways to phase out fossil fuels. But without having a real alternative in place, phasing out fossil fuels will be impossible. What is more, it is not a given that renewable energy technologies suddenly emerge in an energy industry dominated by large fossil fuel–based technologies. In the early seventeenth century in Great Britain, coal was providing much less energy than traditional sources such as human, horse, wood, wind and water provided (Wrigley 2013). However, coal was growing much faster than traditional energy sources, and towards the end of the nineteenth century, coal supplied more than nine-tenths of all the energy consumed (Wrigley 2013). Although the steam engine was in commercial use in the 1780s, it was not until around 1830 that the steam engine became the primary source of power throughout Britain's economy (Goldstone 2006). Change happened slowly, but society and technology co-evolved to the extent that what may have seemed like a trivial new technology, suddenly became an integral part of life. If we take end of 2018 as a point of departure in China, solar and wind power production increased by roughly 50 and 20 percent, respectively, compared to 2017, whilst power generated from coal, gas and oil grew with 7 percent (chinaenergyportal.org 2019b). As shown in Figure 1.1, wind and solar PV installations in China have also grown quickly compared to the rest of the world since 2010. This book provides a detailed comparison of the emergence of China's wind and solar PV industries and attempts to understand better the process within which these industries have grown and what this means in terms of a low carbon energy transition.

The theoretical point of departure in this book is the understanding that technology and society make up two sides of the same coin, and that technological development is a product of the interactions and dynamics between a heterogeneous set of actors, technologies and objects (Bijker, Hughes and Pinch 1987; Sørensen and Williams 2002). A technology is never 'purely' technological: it always is intertwined with society, and vice versa. In general, this implies a certain curiosity as to who and what may be regarded as important or not within innovation processes. It is increasingly acknowledged that technological and industrial development do not take place in closed geographical systems but are the result of cross-border encounters and the mutual learning processes of both international and local actors (Hansen

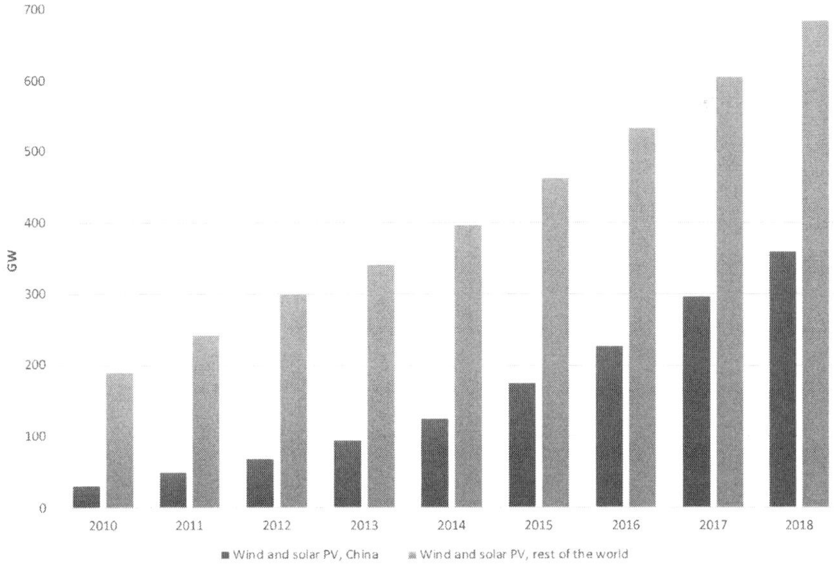

Figure 1.1 Installed capacity of wind and solar PV in China and the rest of the world between 2010 and 2018.

Sources: IRENA (2019).

2009; Nahm and Steinfeld 2014). This is especially evident in China, where foreign technology has contributed substantially to industry development over the past 40 years, and it is also the case for China's wind and solar industries. Who are the actors who strategically work to build wind and solar technology in China, and what type of knowledge and experience do they make use of? This book critiques the way in which technological learning in latecomer countries has been conceptualised through 'technology transfer' and 'catch-up'. What is needed instead is a critical view of how innovation and technology development can be understood in a Chinese context. By introducing and discussing two different perspectives on innovation and social change, sustainability transitions (Markard, Raven and Truffer 2012) and Actor-Network Theory (ANT) (Latour 2005), I intend to bring more detail to the discussion of what innovation can mean—seen from the point of view of the wind and solar industries. The interesting debate which these two perspectives generate is that whilst the sustainability transitions literature has a rather fixed idea of what the essential ingredients of innovation are, ANT questions the assumption that there exists a blueprint which in any certain way can point towards innovation. My ambition when introducing these different perspectives is not to create a unified theory of innovation, but rather to understand innovation processes by employing different lenses that, together, may assist in providing a richer description of the development of wind and solar technologies in China, in this case, the wind energy and solar energy transition. I will return to this discussion in the next chapter.

Over the past decade and a half, the world has seen an explosion of renewable energy installations, and this development has been accelerated and led by China (Yang *et al.* 2016). China is today the country with the most installed bio power, wind power, hydropower and solar PV (REN21 2018), and China is also world leading in manufacturing several of these technologies. Under the 2015 Paris Agreement, China promised to develop renewable energy capacities by 2030 equivalent to the entire US electricity system in 2017 (Roberts 2017). The large amount of GHG emissions coming from China combined with the burgeoning renewable energy industry makes China the single most important country in terms of global GHG mitigation. The nation's almost insatiable electricity consumption is also one of the reasons why China is the world's largest investor in new renewable energy industries (FS-UNEP 2018), such as the wind and solar industries. As will be elaborated on in later chapters, this development is also motivated by energy security concerns, industry development and mitigating pollution and climate change (Ydersbond and Korsnes 2016).

Globally, renewable energy sources are still dependent on policy support in order to compete with traditional energy sources and their growth hinges on continued political prioritisation (REN21 2018). There has been a dramatic reduction in wind and solar PV power prices making renewable electricity cheaper than fossil- or nuclear-fuelled electricity in many parts of the world (REN21 2018). Still, fossil fuels are subsidised and favoured in many areas (REN21 2018). The success of renewable energy industries therefore gives the government a more direct role. In China, there is a large and growing base of renewable energy technologies. From these growing industries, several insights can be gained into the Chinese Communist Party's (CCP) attempts to orchestrate the development. The story of China's remarkable renewable energy development process starts in the 1980s and '90s with an emerging interest in using wind and sun energy for off-grid electricity production in rural areas. From the wind concession programme launched in 2003, China's wind industry took off and grew to an installed capacity of 207 GW at the end of 2018 (GWEC 2019a).[4] In a few years, China conditioned the growth of wind turbine manufacturing and deployment at a scale and pace never before witnessed. Today, the Chinese wind industry produces more wind turbines than any other country, Chinese companies dominate the list of the world's largest wind turbine manufacturers and China has the largest installed capacity of wind energy globally (REN21 2018). For solar PV, the growth of domestic installations has been even quicker. Until 2008, China had installed a total of only 0.14 GW of solar PV whilst it was already a global champion of PV panel manufacturing (PVPS 2017). This means that China was exporting most of the PV panels—something which changed abruptly around 2009 with the financial crisis. By the end of 2018, China achieved a total installed capacity of 174-GW solar PV (chinaenergyportal.org 2019b)—an explosive growth that will be explored and explained further in Chapters 4, 5 and 7. Today, China is world leading in solar PV manufacturing output and installed capacity. Figure 1.2 shows the development of domestic installed capacity of wind and solar PV since 2003.

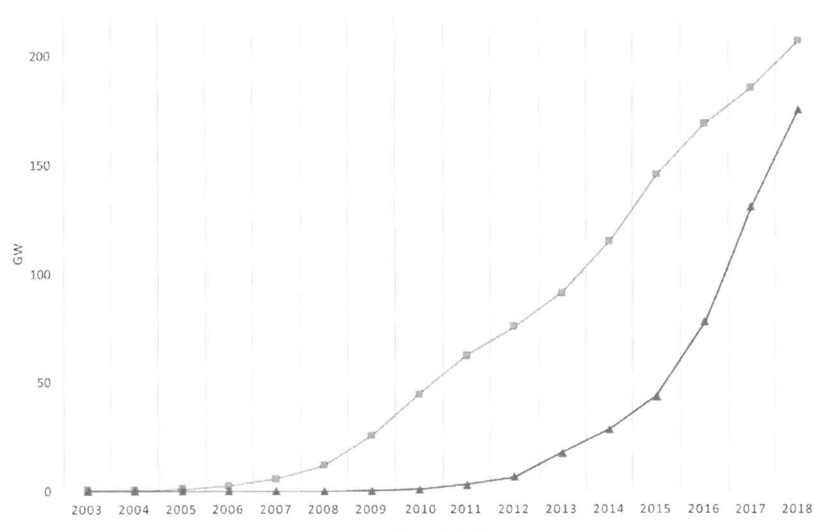

Figure 1.2 Installed wind and solar PV power capacity in China in GW, 2003–18.

Sources: GWEC (2019b) for wind installations and IRENA (2019) for solar PV.

China has showed an increasing focus on attempting to become more independent from foreign firms. In practical terms, as Chinese firms design and master new technologies which become global innovations, they can 'collect rather than pay royalties and license fees' (Kroeber 2011, p. 63). This strategy was, during the Hu Jintao and Wen Jiabao leaderships (2002–12), manifested in what was called the 'indigenous innovation' programme, which, amongst other things, involved financial incentives for developing high-technology products, building a larger base of national intellectual property and standards, and compelling foreign companies to reveal their commercial secrets (Kennedy 2010; Cao, Suttmeier and Simon 2009). The strategy was formulated as a response to the past 20 years of development, which yielded high economic growth, but was disappointing in other respects: low innovation in commercial technologies, lower gains due to royalties paid for foreign technology, devastating environmental impacts, and challenges of meeting national defence needs (Cao, Suttmeier and Simon 2009). Also, after China's entrance into the World Trade Organisation (WTO) in 2001, the nation had to give up some of its policy tools, which had previously been useful in ensuring technology transfer from foreign companies investing in China (ibid.). Therefore, ensuring innovation and technology development domestically has been a core concern. This is also true in the renewable energy industry, however, as pointed out in this book, both the wind and solar industries are facing particular challenges which will need to be solved if China is to succeed.

As Xi Jinping and Li Keqiang came to power the 'indigenous innovation' strategy has been complemented by the 'Made in China 2025' programme, a ten-year strategy which is not only focussing on innovation in high-tech sectors, but also in more traditional and service sectors as well. According to Economy (2018, p. 119), 'the strategy includes localizing and indigenizing technologies and brands, substituting foreign technologies, and capturing global market share'. I will return to a discussion of this strategy in Chapter 3. For now, I want to point out the conundrum which these strategies entail: reducing technological dependency does not appear to reconcile with China's obvious interests in attracting and learning new technology and experiences. What is more, the 'indigenizing' strategy is at odds with the very global nature of these two industries. By studying the interactions between a heterogeneous set of wind and solar PV industry actors from inside and outside China, this book aims at understanding who the protagonists and antagonists are in innovation processes in China's wind and solar industries.

Several China scholars have noticed that after Xi Jinping became chairman of the party in 2012, power has become more concentrated with the CCP, and the role of the state appears to be strengthened (Bøckman 2019; Lardy 2019). According to Lardy (2019), the slowdown in economic growth in China occurred due to a turn away from market-oriented reform after 2012, accompanied with an increase of the influence of state-owned enterprises (SOEs) on domestic economy (Lardy 2019). This could well be the case, but judging by the growth of renewables in the same period, the change is not so evident. This may, on the one hand, mean that we need to be modest about what the analysis of the wind and solar energy industries say about Chinese economy and politics in general. On the other hand, it could also mean that a part of Xi's strategy of reduced growth, resulting (as Lardy argues) from more state control, is driven by an actual interest in increased environmental sustainability. This would also resonate with Xi's stated interest in an 'ecological civilisation' and clean energy (Finamore 2018), increased focus on quality as opposed to quantity (Kirkegaard 2019), and the move from manufacturing to innovation. Such thoughts will be explored more in detail throughout the book. If there is one thing we can learn from the renewable energy industry in China, it is perhaps that we should not underestimate the influence of the Chinese government, as well as the means and tools it has to carry out its long-term plans and strategies.

The advantages of a domestic wind turbine and solar PV manufacturing industry were perceived to be numerous. In addition to the 'indigenous innovation' concern, the Chinese government's goal from the outset was not only to accelerate the development of wind farms, but also to create a cheaper option than reliance on imported turbines (Howell *et al.* 2010; BNEF 2010). In general, China has a lot of experience in manufacturing; therefore, it seemed perfectly natural for the nation to initiate the wind industry, as China has attracted industry after industry since the beginning of the reform period in 1978 (Kroeber 2011). However, as we will get back to in Chapters 4 and 5, the solar PV industry did not receive much central government support at the outset, making the comparison between these two industries especially interesting. As the Chinese solar PV manufacturing industry started taking off, the domestic central and provincial governments also realised its importance and started increasing support. By the end of 2017, about 510,000 people were employed in China's wind industry (44 percent of the world total) and a staggering 2.2 million people were employed in jobs related to manufacturing and installation of solar PV (66 percent of the world total) (IRENA 2018). In 2010, China's renewable energy industry (including solar, wind and biomass energy) was named one of seven new strategic emerging industries the Chinese government named to help advance economic development, foster innovation and promote domestic technologies (CGTI 2011). According to this strategy, an estimated US$231 billion would be invested exclusively in wind power between 2011 and 2020 (Pan, Ma and Zhang 2011). Indeed, renewable energy industries are highlighted as part of three central government industrial-policy documents: The strategic emerging industries catalogue, the 'Made in China 2025', and the 13th Five-Year Plan (Kenderdine 2017). A much stronger government support signal than this is hard to find in China; undoubtedly, renewable energy industries are of strategic importance to China (Christensen 2013; Nahm 2014).

China is heavily reliant on coal as its main fuel to generate electricity, which is connected to the fact that coal is relatively easily available. China has about 13 percent of world coal reserves, and about 1.1 percent and 1.8 percent, respectively, of oil and natural gas reserves (Yi-chong 2017, p. 30).

These reserves are depleting fast. The reserves-to-production ratio, which shows the remaining amount of the resource measured in time, is only 35 years for coal (Yi-chong 2017). China produced 46 percent of the world total of coal in 2014 (Yi-chong 2017). China is still also providing various types of direct and indirect subsidies to coal power, giving it a preference to renewable energy sources (Yuan *et al.* 2019). What is more, there is a mismatch between where energy is demanded, and where the resources are located in the country. Coal and water are the two main fuels for electricity generation, and these resources are concentrated in areas where the population density and consumption levels are relatively low (Yi-chong 2017). The same can be said for the onshore wind and the solar resource, which is concentrated in northern and western parts on China. This means that energy needs to be transported over large distances, either in the shape of coal, or in the shape of high-voltage electricity. In fact, in 2011, about 50 percent of all rail transportation in China was for transporting coal (Yi-chong 2017).

In general, electricity grid companies take on a doubly challenging role. On the one hand they represent the established electricity regime and have sunk costs and vested interests in that system (Unruh 2000). On the other hand, they are necessary in order to transport the electricity from new renewable technology. As you often hear representatives from grid companies say: 'There's no transition without transmission'. In Germany, for instance, large grid-utilities were initially fiercely against the introduction of more renewables and used several means to combat the progressive German Feed-in tariff policy (Jacobsson and Lauber 2006). In China, there are only two electricity-grid companies which manage the whole national power-grid system: State Grid Corporation of China ('State Grid') and Southern Grid. The power grid is separated into regional clusters, which are operated rather independently (IEA 2011). The most important grid company for the wind farms and solar PV plants is State Grid, which operates four of the six regional grid clusters, mainly in the north of China (IEA 2011). As we will get back to, managing such a large grid system is a huge task, particularly complicated by the introduction of electricity, which is produced only when the sun shines and the wind blows.

Since the 1990s, the international community has increased its pressure on China to reduce GHG emissions, especially after 2007 when it became clear that China was the world's largest emitter of CO_2 (Economy 2010). Even though China has rapidly established a renewable energy industry, competing industries internationally are not entirely happy to observe this rapid growth. The growth of Chinese companies in both the wind turbine and solar PV industries have led to increased competition internationally as well as rapidly declining prices of these technologies. Indeed, Green (2019, p. 633) points out that 'without this manufacturing shift to China and strong US investor support, photovoltaics would likely have remained promising, but still too expensive for widespread uptake'. The solar industry was export-oriented and, after 2009, increasingly focused on the domestic market. The wind industry has mainly been domestically oriented but seeks to expand internationally—especially to markets where European competitors are not present, such as in African countries (Economy 2018). China's presence in international markets has led to declining market shares and indeed bankruptcies of international wind and solar PV firms, and industry associations in places like the United States and Germany blame China. It is therefore possible to argue that China is being criticised for not doing enough for the climate, and simultaneously is being criticised for doing too much. This is especially the case, given that China is producing a large share of the world's products which are then exported—leaving most of the carbon footprint in China. This book will not delve deeper into this discussion, but it serves as a backdrop to show some of the complexities within which the growth of China's wind and solar PV industries are tangled.

The chapters of the book address several different aspects of the multifaceted road from having virtually no renewable energy technology in the early 2000s to becoming world leading in manufacturing and power generation of wind and solar PV in the end of 2018. The next chapter presents the theoretical approach of this book and explains the way in which transition and innovation can be conceptualised and understood. The theoretical perspectives employed in this book highlight different aspects of innovation, from technology transfer and a conceptualisation of technological learning in 'latecomer' countries such as China, to more institution- and action-oriented perspectives of innovation. As already mentioned, these approaches may help us comprehend how innovation can be conceptualised and understood in China's wind and solar PV industries. Chapter 3 elaborates on China's current situation in terms of energy governance and the way in which innovation has been understood and attempted catered for. Managing the energy sector for a country with 1.4 billion people is no simple matter. The chapter looks into the various ways in which the central, provincial and local governments attempt to ensure that electricity is provided for the whole population. The chapter points to the fragmented nature of this governance and describes the role of large SOEs in the energy sector, such as State Grid. The chapter also describes the effort which the central government is making to ensure innovation, transforming China from a manufacturing base to a design and innovation power.

Chapter 4 addresses the dynamics of change within China's energy sector and shows the different ways in which the wind and solar PV industries came about. The chapter illustrates that flexible governance has allowed for rapid growth in periods whilst also ensuring that overcapacity does not lead to collapse. Fragmentation and centralisation are identified as important features of the flexibility which the Chinese government has drawn on. Fragmented authority means that there are several concentrations of power in China, such as various central government ministries, local and provincial governments, SOEs and private firms. Centralisation occurs when the very apex of the CCP, typically the standing committee (consisting of the seven most influential people in the party), intervenes directly to make a decision. The chapter contributes with a better understanding of the mechanisms of policy learning and the workings of fragmentation which have laid the foundations for China's wind and solar PV industries. The chapter highlights the important role of policy experimentation in the coordination of interests as part of the policy learning process.

Chapter 5 addresses the international relations of China's wind and solar PV industries. The chapter describes the different ways in which ties between Chinese and international actors have been shaped and enabled the domestic Chinese industries to grow. The chapter also looks more specifically at the role of foreign companies in China's offshore wind industry—in particular the role of European certification and advisory firms. Certification and advisory firms have an increasingly important role in innovation and development processes. As risk and uncertainty become fields to be managed, certification and advisory firms arguably gain an important role by providing evidence that a technology will function within a set of given conditions. European certification and advisory firms played a role in the development of offshore wind in Europe. Further, certification and advisory firms have proven to be important 'intermediaries' in technology development and transfer settings. Thus, by better understanding the roles and strategies of European certification and advisory firms in the development of offshore wind energy in China, the chapter shows how these European intermediaries impact innovation and technology development in China. The chapter points out that, although knowledge about best practice is useful, a certificate from a foreign company appears to be neither required nor desired in the domestic Chinese market. Differences in risk perception, views on the service industry and a lack of attention to local needs complicate matters for entering certification and advisory firms. The chapter underlines the different evaluations of risk between China and Europe which mean that certification and advisory agencies are not similarly acknowledged as legitimate actors in both regions. However, this does not rule out the possibility that certification and advisory agencies will be seen as necessary at later stages in the technology development process. The likelihood that such companies will be of foreign origin may, however, not be large.

Chapter 6 investigates the way in which future-orientation is drawn on as a resource for change in China's offshore wind industry. The success or failure of an emerging technology is contingent on peoples' claims and thoughts about the future and experiences of the past (Einsiedel 2009). An emerging technology is therefore dependent on and influenced by the expectations and visions of the surrounding society, media, government and scientific community. Expectations and visions have proven to be important in creating a 'protective space' or 'niche' for new technologies which may grow to become important technologies (e.g., Geels and Raven 2006). As China develops offshore wind energy, it is interesting to find out what the role of expectations and the future has been there. The chapter finds that ambitiousness about the future, but ambiguousness in implementation, is a strategy employed by the government to ensure change. Founded on a plan-centred economic history, China's visions border on loose plans, and the dynamic between imaginaries and expectations can be characterised as an efficient planning tool used by the government to calibrate plans for the future. Government targets which 'fail' to be reached are therefore not failures per se, but important indicators which are used to calibrate new targets. By making a novel nuance between expectations and imaginaries, the chapter contributes to a better understanding of the dynamics of expectations and their constitutive force in China: as industry actors typically expect something to happen, the negotiation centres on *the extent to which* government imaginaries could and should be performative.

Chapter 7 covers a topic which is hitherto little studied and very new: the role of rooftop and ground-mounted PV systems in urban and rural areas of China, where the electricity is consumed locally (referred to here as 'distributed PV'). The chapter goes into detail investigating the reasons for increased use of urban and rural distributed PV in China, as well as challenges this development entails for local electricity-grid networks. Using a bottom-up perspective, the chapter finds that the use of distributed solar PV systems appears to be disconnected to the 'smart' and 'eco' city projects in China. Three reasons for this disconnect were identified. First, 'smart' and 'eco' city projects have mainly been focussing on economic growth, and issues relating to lacking communication between ministries and government departments hampered connections to solar PV projects. Second, the reluctant role of State Grid in reforming the electricity market and to sell electricity back to the grid complicated the connection with the 'smart' agenda. And third, the increasing use of distributed PV in more urban settings happened as a response to the lower availability and profitability for centralised utility-scale PV projects, not as a coordinated 'smart' or 'eco' effort. This chapter describes some of the challenges which will be exacerbated with the planned deregulation of China's electricity system in 2020, which is highly central to the success or failure of a continued energy transition in China.

Chapter 8 concludes the book by reviewing and discussing some of the core findings. The chapter summarises some of the theoretical contributions of the book and synthesises the way in which innovation can be understood in China's wind and solar industries. The chapter provides an overview of Chinese institutions of learning, as seen through studying the wind and solar industries, expands on the important role of experimentation, and further develops the concept of 'the arts of the state'. Chapter 8 also takes the opportunity to discuss the sustainability of China's energy transition, as well as the implications this transition has for China and the world.

Methodological considerations

This book is based on data collected in Beijing between August and December 2011 (focusing on wind and solar PV), in Shanghai between May 2013 and April 2014 (focusing on offshore wind), and in Beijing between April and July 2018 (focussing on distributed PV), as well as interviews in Norway in early 2013 and 2019. The data mainly consists of interviews, but it also includes a two-month period of participant observation in Shanghai, several field visits, as well as attendance at several industry conferences, exhibitions and workshops in China and Norway between 2011 and 2018, through which I had countless conversations about the topic. During many of these events I took notes or recorded memos in order to remember what I learned from the many conversations and presentations. The interview data were analysed according to principles suggested by 'abductive reasoning', meaning that both field data and existing theory were allowed to influence the researcher, but, ultimately, instead of forcing either the theory or the data into a framework, the researcher left space for his or her own logical reasoning (Reichertz 2007). The analysis was also inspired by constructivist grounded theory, implying that the analysis started already at the data collection stage, and interview structures changed as a better grasp of the topics was developed (Charmaz, 2006; Clarke, 2005). This enabled a greater focus on topics of interest throughout the data collection, and it made analysis easier, as all the data had been gathered.

According to Reichertz (2007, p. 216) 'abduction is intended to help social research, or rather social researchers, to be able to make new discoveries in a logically and methodologically ordered way'. Abduction does not imply a simple combination of fieldwork data and existing knowledge in the field; rather, abduction is a process which occurs when there is no appropriate explanation for an occurred phenomenon. This process happens inside the mind of the researcher, and it is therefore a 'cognitive logic of discovery' (Reichartz 2007, p. 220). As Reichertz (2007, p. 221) puts it:

> Abductive reasoning is [...] an attitude towards data and towards one's own knowledge: data are to be taken seriously, and the validity of previously developed knowledge is to be queried. It is *a state of preparedness to be taken unprepared*. (Emphasis added)

The researcher must make an effort to stay open and not conclude or shape a theory too early. Put differently, the researcher must avoid premature 'emplotment' (Kirkegaard 2015) and make sure not to enrol in the metaphors of those who have the power to create them (Star 1991). I have tried to follow this advice when working with the data. The approach taken here has the advantage of addressing Chinese policymaking from a specific vantage-point: namely the renewable energy industries. In this respect, the method allows for an analysis which casts a fresh perspective on policymaking which is not as strongly influenced by western Chinese scholarship, which has been critiqued of focusing either too much on a 'transition from authoritarianism to democracy' perspective, or of an 'authoritarian resilience' perspective (Shue and Thornton 2017). This is also in line with following the actors and trace the ways in which an actor-network can emerge and stabilise through a co-creation process of negotiations and translations.

In China, access to information and relevant interviewees can depend on the sensitivity of the research topic. Heimer and Thøgersen (2006, p. 15) point out that political sensitivity in China 'becomes a question of timing as well as the audience and the use of the information'. Having conferred with experienced researchers in the field of energy and China before I entered the field, I established that renewable energy is not a sensitive topic compared to issues relating, for instance, to nuclear power. The renewable energy industry appeared to be a field China wanted to display widely, and it was not as difficult as expected to locate and contact relevant people. Interviewees generally appeared to enjoy talking about their work and visions of China's sustainable future, and they appeared to be genuinely interested in answering my questions. In general, language was the largest barrier for both the fieldwork and the written information I accessed, although I received great assistance from my interpreters. The wind and solar PV industries in China are international, and many of the central actors spoke English well. Furthermore, a large number of foreigners engaged with environment, energy and business topics are situated in China, all of whom communicated with me in English. Although many of the interviews were conducted in English, several interviewees were not native English speakers. Use of a foreign language may cause people to unintentionally say things they do not really mean or may tweak the meaning of what was originally intended. This was dealt with by repeating the question or the meanings when there was any doubt, in order to confirm. I am therefore confident that the information gathered through the interviews and used in this book represents the interviewees' meanings. I also found it convenient to engage in several classes of intensive language training whilst in China. I thus acquired a basic knowledge of Mandarin, which was very useful in establishing a positive interview atmosphere, as it conveyed my genuine interest.

The interview situation has received much attention in qualitative research methodology because of the central position of the interviewer in a classical subject-object positivist discussion. In a constructivist approach, the focus is on the interaction between the interviewer and the interviewee. In this interaction, each party's perceptions of the other influence the outcome of the interview, and the outcome, in itself, is a dialogue, rather than a set of facts retrieved from a source (i.e., there is a co-construction of meaning). During the 2011 6-month fieldwork in China, I conducted 12 interviews; during the 2013–14 12-month fieldwork, I conducted 43 interviews; and during the 2018 3-month fieldwork, I conducted 21 interviews. I also did some interviews back in Norway and over the phone. An overview with dates, location and background of these interviews can be found in the appendix. The interviews which are directly referred to in the text are numbered and marked according to the list provided in the appendix of this book. In total, 77 interviews have informed the present analysis. Most interviews lasted around one hour. All interviews were semi-structured, containing 'a sequence of themes to be covered, as well as suggested questions' (Kvale 1996, p. 124).

An interview guide was used flexibly, allowing for a change of theme which naturally corresponded with an interviewee's reasoning, and answers were followed up as appropriate. Some interviewees had prepared PowerPoint presentations, and it was sometimes difficult to explain that I was more interested in their opinions about the wind and solar industries than facts and figures. The information they provided was nevertheless useful as a point of departure to discuss concrete issues which were more interpretive. Topics which were referred to as established 'facts', such as 'the government target for 2015 is 5GW', could therefore launch discussion of the likelihood of reaching the target, the ambition of the government versus the industry and similar topics. Interviews always began with me going through the purpose of the study, then clarifying the amount of time needed to complete the interview, as well as my plans for using the information gathered; finally, I informed the interviewee of my policies on confidentiality and anonymity (Creswell 2007). All interviewees are anonymous in order to protect their identity. Upon request, some of the interviewees received a copy of the interview guide before the interview.

Collecting data for the study of distributed solar PV in 2018, I made three different field visits: One two-day stay at a small village in Shandong Province, and two visits to large rooftop PV installations in Beijing. The village was called Huairen and had a population of about 10,000 people. I went to this village together with four other students from Tsinghua University, who also were collecting data to do research on households in this village. We visited ten households, à 5.3-kW rooftop PV systems, all had solar hot water (SHW), and we also visited a large local PV installation of 40 megawatts (MW). In Beijing I visited two different locations with rooftop PV. Location 1 was an indoor ski resort facility with 300-kilowatt-peak (kWp) rooftop PV. At the site I interviewed a local manager of the company which was responsible for installation. Location 2 was a design and research institute facility with 500-kWp rooftop PV. At the site I interviewed the local project engineer and a division manager of the institute. This data will be presented in more detail in Chapter 7.

I employed several different interpreters during the whole stay in China: two in Beijing, two mainly around Shanghai, and one near Guangzhou. Three of them were professional interpreters with previous experience from renewable energy industries, and two were master's students. Use of an interpreter has both advantages and disadvantages: it allows for flexibility regarding note-taking during interviews, but also heightens the probability of information manipulation in the reiteration (Scheyvens and Storey 2003). Most of my translators had worked for several years as professional interpreters and had held assignments within the wind and solar industries. Based on the interpreters' experiences and professionalism (e.g., using industry-specific words), I was assured that the interviews would be interpreted as accurately as possible.

On the one hand, it appeared easier to get access to the right people as a foreigner in China, because of the possibility that I might represent some important foreign organisation. This was pointed out to me by several of my Chinese friends before, during and after my stays in China. For example, one Chinese interviewee said that he was interested in contributing to my research projects because he was interested in learning more about foreigners and what topics we were interested in. On the other hand, people might not have told me how things really are. This concern is naturally prevalent in any kind of interview-based research, but in this case it might be different for two reasons. First, we can imagine that interviewees might attempt to paint a more 'beautiful picture' because I am a foreigner, or a Westerner, and Chinese might want to compare favourably to Westerners. Second, because saying something else would interfere with the official CCP policy. As pointed out by Solinger (2006), interviewing can sometimes be challenging in China, as Chinese respondents are careful about saying something wrong or out of line with official policy. Thus, although the renewable energy field itself appeared to be a relatively insensitive topic, it was clear that top-level politics typically were not up for discussion. Several interviewees answered in a general and abstract way and left room for interpretation in their responses. Some of these 'hints' and vague statements from Chinese interviewees could therefore be regarded as fairly strong indicators of opinions, though they might not sound like it. Some such remarks may also have been due to a lack of mastery of English, in the cases where no interpreter was used.

I attempted to deal with this challenge by asking questions in various ways, such as repeating the interviewee's response and asking for confirmation. Some interviewees were also worried that I would use the information in news articles or similar outlets. In such situations, it helped to reassure them that the information gathered was only for research purposes, and that it would be made strictly anonymous. Also, it was often relatively obvious if we were touching on a sensitive topic (for instance, when talking about opinions of government policy), and I would change my questions accordingly. In general, my main method of making sure that I had relatively reliable data was to compare and contrast, or triangulate, several sources of information. These sources were written information from inside and outside of China, such as research articles, books, or online news, as well as interview data, and data from my participant observations and observations at different wind- and solar-related venues. Nevertheless, I was not able to use information written in Chinese, and this is a limitation to the gathered information. Lastly, the propaganda apparatus of the CCP may have had a reach that went beyond any of my attempts to get reliable data. In this respect, however, it is interesting to note that the people who were the most outspoken and critical to the way things were developing typically had a high-level position within either the government or state-owned enterprises. Such interviewees provided very interesting perspectives which would balance accounts from other sources. Hence, ultimately, my own intuition of what was a reliable or a less reliable account would also influence my analysis of the data.

You might also ask how it is possible to address large entities such as 'the government', the CCP and European certification agencies, and still claim to have had a 'grounded' approach. In an attempt to move away from preconceived levels and hierarchies, Clarke (2005) understands each situation as constitutive of its own levels. In other words, I analysed these larger entities as they appeared inside the actual situation under examination. Callon and Latour (1981, p. 299) formulated it in this way:

> A macro-actor is a micro-actor seated on black boxes, a force capable of associating so many other forces that it acts like a 'single man'. The result is that a macro-actor is by definition no more difficult to examine than a micro-actor.

The various actors I studied defined their worlds through their own concepts, and my interpretation of the data was shaped by interviewee perceptions that, in turn, formed my hypotheses. In this way, it was possible for me to study macro-processes within an international industry without referring to preconceived levels. The interviewer and the interviewee both refer to and conceive of these levels, in situ. Levels and hierarchies can therefore be understood as being constructed by the actors involved, and not being preexisting frames.

Notes

1 Note that when talking about energy this book focuses mainly on the electricity sector, and not on heating. When referring to solar energy, the book focuses on solar photovoltaic systems, and not solar hot water or concentrated solar power.
2 For an excellent discussion on this in a Chinese context, see Tyfield (2017).
3 Nuclear energy technologies are also considered low carbon, and China is indeed investing a lot in this technology as well. By 2020, China will have almost 50 nuclear power bases (Kenderdine 2017). This book will nevertheless focus on wind and solar PV technologies.
4 Note that the Global Wind Energy Council (GWEC) appears to have higher numbers of installed wind capacity than China's National Energy Administration (NEA). The reason for this is most likely that the NEA only counts capacity which has been connected to the grid by the end of the year, whilst GWEC reports all installed capacity.

References

Andrews-Speed, P. and Zhang, S., 2019. *China as a Global Clean Energy Champion. Lifting the Veil*. Singapore: Palgrave Macmillan.

Alva, C.A.H. and Li, X., 2018. Power sector reform in China. An international perspective. Report published by the OECD/IEA, International Energy Agency.

Bijker, W.E., Hughes, T.P. and Pinch, T., 1987. *The Social Construction of Technological Systems. New Directions in the Sociology and History of Technology*. Cambridge, Massachusetts: MIT Press.

BNEF, 2010. *Joined at the Hip: The US–China Clean Energy Relationship*. Bloomberg New Energy Finance (BNEF). Available at: www.bnef.com/WhitePapers/download/27. (Accessed 7 August 2011).

Bøckman, H., 2019. Under Xi er Kina blitt stadig mer en partistat. Norwegian Bokmal. Available at: www.aftenposten.no/meninger/kronikk/i/b5QQn5/Under-Xi-er-Kina-blitt-stadig-mer-en-partistat-Harald-Bockman (Accessed 22 May 2019).

Callon, M. and Latour, B., 1981. Unscrewing the big Leviathan: how actors macro-structure reality and how sociologists help them to do so. *Advances in Social Theory and Methodology: Toward an Integration of Micro and Macro-Sociologies*, pp. 277–303.

Cao, C., Suttmeier, R.P. and Simon, D.F., 2009. Chapter 10. Success in state-directed innovation? Perspectives on China's plan for the development of science and technology. In G. Parayil and A.P. D'Costa (eds). *The New Asian Innovation Dynamics: China and India in Perspective*. Houndmills: Palgrave Macmillan.

CGTI, 2011. *The China Greentech Report*. Beijing: The China Greentech Initiative (CGTI). Available at: www.china-greentech.com/report.

Charmaz, K., 2006. *Constructing Grounded Theory: A Practical Guide Through Qualitative Analysis*. Thousand Oaks, London, New Delhi: Sage Publications.

chinaenergyportal.org, 2019a. Notice on the first batch of 2019 of non-subsidized wind and PV power generation projects (grid-parity projects). Available at: https://chinaenergyportal.org/en/notice-on-the-first-batch-of-2019-of-non-subsidized-wind-and-pv-power-generation-projects-grid-parity-projects/ (Accessed 23 May 2019).

chinaenergyportal.org, 2019b. 2018 electricity & other energy statistics. Available at: https://chinaenergyportal.org/en/2018-electricity-other-energy-statistics/(Accessed 27 May 2019).

Christensen, N.H., 2013. *Shaping Markets: A Neoinstitutional Analysis of the Emerging Organizational Field of Renewable Energy in China*. PhD thesis at the Doctoral School of Organisation and Management Studies at Copenhagen Business School. Available at: http://openarchive.cbs.dk/bitstream/handle/10398/8627/Nis_Høyrup_Christensen.pdf.

Clarke, A.E., 2005. *Situational Analysis: Grounded Theory After the Postmodern Turn*. Thousand Oaks, London and New Delhi: Sage Publications.

Creswell, J.W., 2007. *Qualitative Inquiry & Research Design: Choosing Among Five Approaches* (2nd ed.). Thousand Oaks, CA: Sage Publications.

Economy, E., 2010. *The River Runs Black: The Environmental Challenge to China's Future* (2nd ed.). Ithaca, NY: Cornell University Press.

Economy, E., 2018. *The Third Revolution. Xi Jinping and the New Chinese State*. Oxford University Press. Kindle Edition.

Einsiedel, E.F., 2009. *Emerging Technologies: From Hindsight to Foresight*. Vancouver, Toronto: UBC Press.

Finamore, B., 2018. *Will China Save the Planet?* Wiley. Kindle Edition.

FS-UNEP, 2018. Global trends in renewable energy investment 2018. Report prepared by Frankfurt School-UNEP Centre/BNEF. Available at: www.fs-unep-centre.org (Accessed 23 April 2019).

Geels, F. and Raven, R., 2006. Non-linearity and expectations in niche-development trajectories: ups and downs in Dutch biogas development (1973–2003). *Technology Analysis & Strategic Management*, 18(3–4), pp. 375–392.

Green, M.A. 2019. How did solar cells get so cheap? *Joule*, 3, pp. 631–640.

Goldstone, J.A., 2006. Thoughts on the industrial revolution. Available at: www.lse.ac.uk/Economic-History/Assets/Documents/Research/GEHN/GEHNConferences/conf9/Conf9-Goldstone.pdf (Accessed 12 March 2019).

GWEC, 2019a. Global wind report 2018. Global Wind Energy Council (GWEC). Available at: https://gwec.net/global-wind-report-2018/.

GWEC, 2019b. Global wind reports. Global Wind Energy Council (GWEC). Available at: https://gwec.net/members-area-market-intelligence/reports/gwec-global-reports/.

Hansen, G.H., 2009. *Encounters on the Far Side: Internationalization and Domestication of Norwegian Lifeboats in China.* Doctoral dissertation at the Department of Interdisciplinary Studies of Culture, Norwegian University of Science and Technology, Trondheim. Accessed at: http://ntnu.diva-portal.org/smash/get/diva2:201015/FULLTEXT03.pdf.

Heimer, M. and Thøgersen, S., 2006. *Doing Fieldwork in China.* Copenhagen: NIAS Press.

Howell, T., Noellert, W.A., Hume, G., and Wolff, A.W., 2010. China's promotion of the renewable electric power equipment industry: hydro, wind, solar, biomass. Dewey & LeBoeuf LLP for the National Foreign Trade Council. Available at: www.nftc.org/default/Press%20Release/2010/China%20Renewable%20Energy.pdf (Accessed 19 May 2011).

IEA, 2011. *Integration of Renewables: Status and Challenges in China.* International Energy Agency (IEA) Working Paper. Retrieved at www.iea.org/papers/2011/Integration_of_Renewables.pdf (Accessed 5 April 2011).

IRENA, 2018. Renewable energy and jobs. Annual review 2018. Report prepared by the International Renewable Energy Agency (IRENA). Available at www.irena.org/publications/2018/May/Renewable-Energy-and-Jobs-Annual-Review-2018.

IRENA, 2019. Statistics time series. Trends in renewable energy. By the International Renewable Energy Agency (IRENA). Available at: https://irena.org/Statistics/View-Data-by-Topic/Capacity-and-Generation/Statistics-Time-Series (Accessed 22 May 2019).

Jacobsson, S. and Lauber, V., 2006. The politics and policy of energy system transformation—explaining the German diffusion of renewable energy technology. *Energy Policy*, 34(3), pp. 256–276.

Kenderdine, T., 2017. China's industrial policy, strategic emerging industries and space law. *Asia & the Pacific Policy Studies*, 4(2), pp. 325–342.

Kennedy, S., 2010. Indigenous innovation: not as scary as it sounds. *China Economic Quarterly*, 14(3), pp. 15–20.

Kirkegaard, J.K., 2015. *Ambiguous Winds of Change—or Fighting against Windmills in Chinese Wind Power. A Constructivist Inquiry into China's Pragmatics of Green Marketisation. Mapping Controversies Over a Potential Turn to Quality in Chinese Wind Power.* PhD Monography, Department of Business and Politics, Copenhagen Business School. Available at: http://openarchive.cbs.dk/handle/10398/9116.

Kirkegaard, J.K., 2019. *Wind Power in China. Ambiguous Winds of Change in China's Energy Market.* Abingdon and New York: Routledge.

Kroeber, A.R., 2011. Chapter 3. Developmental dreams: policy and reality in China's economic reforms. In S. Kennedy (ed.) *Beyond the Middle Kingdom: Comparative Perspectives on China's Capitalist Transformation.* Stanford, CA: Stanford University Press.

Kvale, S., 1996. *InterViews: An Introduction to Qualitative Research Interviewing.* Thousand Oaks, CA: Sage Publications.

Lardy, N.R., 2019. *The State Strikes Back: The End of Economic Reform in China?* Peterson Institute for International Economics.

Latour, B., 2005. *Reassembling the Social. An Introduction to Actor-Network-Theory.* Oxford and New York: Oxford University Press.

Markard, J., Raven, R. and Truffer, B., 2012. Sustainability transitions: an emerging field of research and its prospects. *Research Policy*, 41, pp. 955–967.

Nahm, J.M., 2014. *Varieties of Innovation: The Creation of Wind and Solar Industries in China, Germany, and the United States*. Doctoral dissertation, Department of Political Science, Massachusetts Institute of Technology.

Nahm, J. and Steinfeld, E.S., 2014. Scale-up nation: China's specialization in innovative manufacturing. *World Development*, 54, pp. 288–300.

Pan, J., Ma, H. and Zhang, Y., 2011. *Green Economy and Green Jobs in China: Current Status and Potentials for 2020*. Worldwatch Institute. Available at: www.worldwatch.org/node/8677.

power-technology.com, 2019. China considers how to support solar beyond subsidies. Available at: www.power-technology.com/comment/china-solar-pv/ (Accessed 23 May 2019).

PVPS, 2017. National survey report of photovoltaic applications in China 2017. Report by the International Energy Agency (IEA) Photovoltaic Power Systems Technology Collaboration Programme (IEA-PVPS).

Reichertz, J., 2007. Abduction: the logic of discovery in grounded theory. In A. Bryant and K. Charmaz (eds.) *The Sage Handbook of Grounded Theory*. Sage, pp. 214–228.

REN21, 2018. Renewables 2018 global status report. Renewable Energy Policy Network for the 21st Century (REN21). Available at: www.ren21.net/status-of-renewables/global-status-report/ (Accessed 21 April 2019).

Roberts, D., 2017. By 2020, every Chinese coal plant will be more efficient than every US coal plant. Available at: www.vox.com/energy-and-environment/2017/5/15/15634538/china-coal-cleaner (Accessed 1 June 2019).

Saich, T., 2011. *Governance and Politics of China* (3rd ed.). Houndmills, Basingstoke, Hampshire/New York: Palgrave Macmillan.

Scheyvens, R. and Storey, D., 2003. *Development Fieldwork: A Practical Guide*, London: Sage Publications.

Shapiro, J., 2012. *China's Environmental Challenges*. Kindle Edition: Wiley.

Shue, V. and Thornton, P. (eds), 2017. *To Govern China: Evolving Practices of Power*. Cambridge: Cambridge University Press.

Solinger, D.J., 2006. Interviewing Chinese people: from high-level officials to the unemployed. In M. Heimer and S. Thøgersen (eds) *Doing Fieldwork in China*. Copenhagen: NIAS Press, pp. 153–168.

Star, S.L., 1991. Power, technologies and the phenomenology of conventions: on being allergic to onions. In J. Law (ed.) *A Sociology of Monsters: Essays on Power, Technology and Domination*. London and New York: Routledge, pp. 26–57.

Sørensen, K.H. and Williams, R., 2002. Social shaping on the move? On the policy relevance of the social shaping of technology perspective. In K.H. Sørensen and R. Williams (eds) *Shaping Technology, Guiding Policy: Concepts, Spaces & Tools*, pp. 19–36.

Tyfield, D., 2017. *Liberalism 2.0 and the Rise of China: Global Crisis, Innovation and Urban Mobility*. London and New York: Routledge. Kindle Version.

Unruh, G.C., 2000. Understanding carbon lock-in. *Energy Policy*, 28(12), pp. 817–830.

Wei, Y.M., Chen, H., Chyong, C.K., Kang, J.N., Liao, H. and Tang, B.J., 2018. Economic dispatch savings in the coal-fired power sector: an empirical study of China. *Energy Economics*, 74, pp. 330–342.

Wrigley, E.A., 2013. Energy and the English industrial revolution. *Philosophical Transactions of the Royal Society A: Mathematical, Physical and Engineering Sciences*, 371(20110568), pp. 1–10.

Yang, X.J., Hu, H., Tan, T. and Li, J., 2016. China's renewable energy goals by 2050. *Environmental Development*, Volume 20.

Ydersbond, I.M. and Korsnes, M.S., 2016. What drives investment in wind energy? A comparative study of China and the European Union. *Energy Research & Social Science*, 12, pp. 50–61.

Yi-chong, X., 2017. *Sinews of Power. The Politics of the State Grid Corporation of China.* New York: Oxford University Press.

Yuan, J., Guo, X., Zhang, W., Chen, S., Ai, Y. and Zhao, C., 2019. Deregulation of power generation planning and elimination of coal power subsidy in China. *Utilities Policy*, 57, pp. 1–15.

2 Transition, catch-up and innovation

Different perspectives

Analysing China under the heading of 'transition' is not a new phenomenon. The word 'transition' is a common concept used to explain what has been happening in China since the reform and opening up period that began in 1978 (e.g., Dittmer and Liu 2006; Hussain 2002). Indeed, it may be argued to be an outdated approach, as recently pointed out by Shue and Thornton (2017, p. 3): 'by the dawn of the twenty-first century "transitology", and (…) the "transition paradigm", were already showing signs of having exhausted their usefulness—particularly as "hybrid" political regimes (partly authoritarian and partly democratic) were perceived to be proliferating'. I will return to this discussion in the next chapter. It should be noted here, however, that the present understanding of transition is not referring to a political transition from 'authoritarianism' to 'democracy' nor from 'plan' to 'market', but rather to a sociotechnical energy transition from unsustainable ways of producing and consuming, to more sustainable ones (e.g., Labussière and Nadaï 2018). Clearly, what is considered sustainable or not is also up for debate and is not necessarily given or the same in each country and context. In this book, the understanding of sustainability is mainly related to emissions of greenhouse gases (GHGs) connected to energy production, in particular electricity. The main emphasis is therefore on the supply side, aiming to gain a better understanding of the way in which China can transition from a fossil-fuelled electricity provision, to one more dominated by renewable energies, in particular solar and wind energy.

Technological catch-up

Technological catch-up can be seen as a process in one firm or country of reaching a level of technological sophistication which exists in another 'frontier' firm, country or group of countries. In a long-term historical perspective, technological catch-up can be understood to have been going on since humans started farming, wherein one civilisation became more advanced than another, and this advancement moved as a pendulum between east and west on the Eurasian continent (Goody 1996). Technological catch-up is often measured in terms of the increased productivity and income of a country

(Abramowitz 1986). With this measure, a division is made between countries which are technologically advanced and countries which are technologically 'backwards'. The goal for these 'backwards' countries is to increase productivity and achieve a higher gross domestic product. Stories of technological catch-up typically start with the industrial revolution originating in the United Kingdom in the second half of the eighteenth century, when surrounding countries discovered the advantages of the United Kingdom's technological advances (Bruland and Mowery 2005). The study of technological catch-up has traditionally been the domain of economists and historians studying the way in which European countries such as Germany and France caught up with England (Fagerberg and Godinho 2005; Lundvall 2011).

'Late industrializers', as Amsden (1989) calls them, such as Japan and Korea, often caught up with frontier countries by protecting their industry and cunningly learning to master frontier technology. Based on a historical review of now-developed countries such as Norway, Finland and Japan, Chang (2007) stresses the importance of regulating foreign investments so as to protect and build a national industry. Until the 1980s, for instance, Japan 'was virtually closed to foreign investment' (Chang 2007, p. 16). Interestingly, Japan also emerged as a global technical power in the 1980s concurrent with a more open policy for foreign investment. Thus, opening the economy at the right time appears to be one important aspect of catch-up. Other commonalities of now-developed countries include their considerable cross-border flow of people, active government support for domestic companies and weak regulation of intellectual property rights (IPR) (Nelson 2004a).

IPRs play an important role for the firm which owns the technology, as property rights hinder others from using the technology without paying for it. It has been claimed that strong intellectual property right protection incentivises investment in research and development (Xue and Liang 2010), which is often taken as an indicator of the intensity of a firm's learning effort and, hence, success (Cohen and Levinthal 1990). Originally, developing countries were strongly opposed to an international intellectual property rights system because they argued that it only benefitted states which were already technologically strong (Rosendal 2001). In other words, those who owned the knowledge had the upper hand. Since China became a member of the WTO in 2001 (Suttmeier and Yao 2004), it has had to enforce stronger patent legislation, in line with the trade-related aspects of intellectual property rights (Naughton 2007). Since China has profited from foreign investment, the challenge has been twofold: meeting the demand for strong property right protection of multinational companies and satisfying domestic companies, which 'favored an IPR regime conducive to technology transfer and knowledge diffusion' (Xue and Liang 2010 p. 318). We will get back to different understandings of innovation in a Chinese context in the next chapter.

Technological catch-up can be seen as a rather remarkable feat, given that knowledge is needed to gain new knowledge, skills beget skills and capital fosters more capital (Perez and Soete 1988). For an outsider firm or country, it appears difficult to enter such a complex and self-subsisting system.

However, it is not only the knowledge and experiences of the 'advanced' countries which are important in order to catch up. History has shown that, as countries catch up, they tend to generate innovation and unseen opportunities (Fagerberg and Godinho 2005). Such innovation and unseen opportunities occurred, for instance, when both Germany and the United Statess started catching up with the United Kingdom during the course of the nineteenth century (Fagerberg and Godinho 2005). Perez and Soete (1988) argue that technological catch-up becomes a possibility mainly through 'windows of opportunity' that arise because a new setting creates new opportunities. One such setting is the information and communication technology 'revolution' that started in the 1970s (Perez 2010; Perez 2013). Technological catch-up, understood this way, is distinguished from approaches which see development as a 'race along a fixed track, where catching up will be merely a question of relative speed' (Perez and Soete 1988, p. 460). Instead, catching up is a process wherein a country seeks to reach a certain level of development, and, in that process, may innovate and create new avenues for growth. Consequently, catch-up in this understanding can be seen more as 'creating new' rather than 'imitating old'. This means, then, that in each case of a historical catch-up, whether it was Germany, the United States, Japan or Korea, something new was added. Simplified put, a retrospective look at catch-up may help to gain a better understanding of how a particular country learned, but this knowledge might partly be unusable for another catching-up country as the window of opportunity has already been covered.

Catch-up strategies are not only about *grasping* an opportunity, but also about *creating* new opportunities. As Amsden (1989, p. 329) concludes in her seminal study on technological catch-up in South Korea: 'Japanese and now Korean economic history (…) teaches that we may expect not merely imitation and copying from learners—the longstanding stereotype. We may also expect creativity, because learning itself turns out to be a highly creative process'. Although foreign direct investment can lead to economic growth, catching-up countries typically want to own and master the transferred technology. This is one reason why several innovation scholars refer to learning as the most important source of technological catch-up (Amsden 1989; Lundvall 2007; Lundvall 2011). In other words, successful catch-up boils down to successful learning and innovation. In this learning process—as no country, sector or institution is the same—it appears that adaptability to changing conditions is of the essence: 'Typically, the successful countries have been those that were able skilfully to adapt their policy focus to changing conditions' (Chang 2002, p. 127).

In other words, theories of technology diffusion, transfer and catch-up tend to give too much weight to explanatory factors which come from the outside and should be readily adopted and accepted in the catching-up country. Technology is seen as an exogenous and isolated factor which remains similar across geographical space. This is not satisfactory because it fails to give due attention to the domestic learning processes which are accompanied with the introduction of something new. Hence, we may ask: How can a catching-up country's domestic environment become innovative? This issue is studied closer by evolutionary economists who emphasise the importance of building domestic capabilities which accommodate for new technologies through supporting institutions, organisations and networks—or 'innovation systems' as they are called. Next, we take a closer look at the innovation systems literature in order to see how this line of thought conceptualises innovation and learning, and what it considers significant constituents of industry development.

Innovation systems and learning

The innovation systems literature developed in the 1980s as a response to mainstream economic theory (Lundvall 2010). Innovation systems analysts argue that innovation should not be regarded as a single, isolated event (Kline and Rosenberg 1986). Rather, innovation is seen as ubiquitous, continuous and cumulative—a process whereby interactive learning and collective entrepreneurship are crucial components (Lundvall 1988). Innovation systems theorists take as a starting point the idea that technology develops in a web of interactions between actors, networks and institutions. Innovation is seen to result from a broad spectrum of factors, and not only through market forces or research and development (R&D). Under the innovation systems umbrella, there are several ways of defining levels of analysis, and these can roughly be grouped into two main approaches: one which 'delineates systems on the basis of technological, industrial, or sectoral characteristics' and one which emphasises geography, such as national or regional borders (Fagerberg, Mowery and Nelson 2005, p. 12). Both approaches share a systemic focus, meaning that they do not understand innovation activities in isolation.

National systems of innovation have been defined in different ways. Lundvall (1988) emphasises the interactive aspects of national innovation systems, seeing user-producer relations as important to the innovation performance of a country. These user-producer relationships, Lundvall argues, are crucial for successful innovation and are determined largely by a country's organisational and institutional setup. A common feature in most definitions of national innovation systems is the role given to a country's institutions, and the way in which these institutions interact in order to facilitate innovation (Lundvall *et al.* 2009). Lundvall (2010) draws a distinction between broad and narrow definitions of national innovation systems. The broad definition includes

> all parts and aspects of the economic structure and the institutional set-up affecting learning as well as searching and exploring—the production system, the marketing system and the system of finance present themselves as sub-systems in which learning takes place.
>
> (Lundvall 2010, p. 13)

The narrow definition, on the other hand, includes 'organisations and institutions involved in searching and exploring—such as R&D-departments, technological institutes and universities' (Lundvall 2010, p. 13).

As we can see, institutions have a central role in the story, but what are they, really? Since economic systems are filled with uncertainty, institutions are thought to 'provide agents and collectives with guide-posts for action' (Lundvall 2010, p. 10). These institutions may be 'routines, guiding everyday actions in production, distribution and consumption, but they may also be guide-posts for change' (Lundvall 2010, p. 10). Institutions typically evolve over time and regulate—either directly or indirectly—the way in which actors relate to each other (Johnson 2010). Important features of institutions include their ability to coordinate the use of knowledge, mediate conflicts and provide incentive systems for involved actors (Johnson 2010). Institutions create social stability—stability which in turn produces inertia but also provides the necessary platform for changes to occur (Johnson 2010). Therefore, successful introduction of new technology is determined by the way in which institutions change. Edquist (2005, p. 188) explains that institutions have been treated relatively incoherently within the innovation systems literature, which sometimes refers to the 'rules of the game' and sometimes to the 'players in the game'. For example, Lundvall traditionally referred to institutions as rules, whilst Nelson and Rosenberg treat them as the different kinds of organisations included in an innovation system (Edquist 2005, p. 186). Edquist (2005) therefore distinguishes the two main components of innovation systems into organisations, such as firms, universities and public agencies, and institutions, such as laws, rules and norms.

Learning

Learning is given an important role in innovation systems conceptualisations, and, as established in the previous section, learning is important in the catching-up process of latecomer firms. Learning has been studied by a number of scholars from a range of disciplines, and there is little agreement over what it is and how it occurs (Dodgson 1993). A division between science-based learning and experience-based learning has received increased attention in discussions of learning (Jensen *et al.* 2007; Isaksen and Nilsson 2013; Lundvall 2011). Since policy has traditionally given greater attention to science, research and development as the most important contributions to innovation, Jensen *et al.* (2007) emphasise that both science-based learning *and* experience-based learning are crucial for innovation. Science-based learning is abbreviated as STI (science, technology and innovation) and refers typically to research and development and activities which generate 'know-why'. Experience-based learning is abbreviated as DUI (doing, using and interacting) and refers to more tacit and localised aspects of knowledge which generate 'know-how' and 'know-who'. The DUI learning mode balances the focus on scientific learning, and it is necessary because 'practice in most fields remains only partially understood, and much of engineering design practice involves solutions to problems which professional engineers have learned to "work", without any particularly sophisticated understanding of why' (Nelson 2004b, p. 458). In other words, as we turn to look at China's wind and solar industries, the most important variables which can explain success are not only the amount of research and development, but also more tacit learning processes within firms, and the ways in which actors interact, use and gain hands-on experience in order to steadily improve a product (Kamp, Smits and Andriesse 2004).

One aspect of learning which appears to be somewhat neglected within innovation systems literature is that not only can firms learn, but also other involved actors, such as the customers, local communities or research agencies, can as well. Garud and Karnøe (2003) introduce a different understanding of learning by using the concepts 'bricolage' and 'breakthrough' to describe contrasting learning modes in the Danish and US wind turbine industries. The (successful) Danish version is called 'bricolage', which refers to a path on which the technology and a range of actors are tied together in a process of co-evolution. This mode of learning implies that the designers, manufacturers, users, policymakers and other actors involved in the making of technology must respond to each other and new versions of the technology as it evolves. Learning and agency are distributed across the participating actors; goals are continuously reformulated and technological development proceeds in incremental steps. 'Breakthrough', on the other hand, refers to a path on which certain technological goals are articulated more clearly. The US wind turbine industry pursued a rather unsuccessful 'linear' engineering science-based technology push-model (Garud and Karnøe 2003). Garud and Karnøe (2003) thus conclude that, in some cases, the seemingly simpler bricolage approach may possess significant advantages over the sophisticated high-tech breakthrough approach, as the latter may stifle the micro-learning processes which allow for the mutual co-shaping of emerging technological paths. Indeed, as I will show later in this book, this was important for how policy learning happened in China's wind industry.

The innovation systems approach has also been used to study China. Liu and White (2001, p. 1112), for instance, conclude that

> It is far from clear that evolving into an innovation system similar to that found in developed market economies is a possible or even advisable objective for China or other countries emerging from central planning regimes and Soviet-style industrial organization.

Although it is relatively clear that the Chinese government wants to achieve long-term economic growth and prosperity for a larger share of its inhabitants, we must ask what innovation system may work for China. For instance, how can China ensure that the appropriate types of institutions are developed? Moreover, although institutions and organisations may be present and 'strong', we may also ask who is maintaining the system and who induces change within it. How is such change achieved? One critique of innovation systems approaches may therefore be that it fails to provide tools and perspectives which highlight how institutions, organisations and learning come about and change, and how new ones may emerge.

Somewhat simplified, we may say that if I have been able to map all firms, financing institutions, research institutions and government agencies related to the Chinese wind and solar initiatives, I might know a lot about how the system looks and how one company acquired technology through R&D or other types of learning. Yet, I might have no specific idea of why exactly those actors came into existence, who supported them and what struggles they went through to end up there. Was it a group of five environmentally interested people who started a company, or was it a large state-owned company which decided to start a new and 'green' branch? Was it a farmer who decided to try to harness the wind and got help from his neighbours to build a turbine? Or was it a high-tech, top-down initiative financed by the government? Was it perhaps a large movement of people who were tired of pollution in a Chinese city? In other words, how were these initiatives able to make a mark and create an emerging space for action? These questions highlight movements, initiatives and changing conditions which arise due to larger societal, environmental or political interventions and influences.

Since the internal dynamics of any country or industry arguably is different, it might therefore be too simple to assume that certain organisations and institutions are more important than others. Companies, markets and government policies are different in China, implying that the actions and strategies of involved actors may also be different. Therefore, by centring on a specific technological innovation project, we might neglect societal, environmental, political or other factors which are relevant. In other words, the innovation systems approach fails to adequately address how relations can be seen as both social and technological at the same time (Bijker and Law 1992). In practical terms, this, amongst other things involves examining the co-production of power, conflicts of interest, politics and technology and how this co-production shapes and impacts developments. This may be particularly important in a Chinese context, where the role of the Chinese Communist Party (CCP) is so prevalent and impacts so many processes (Tyfield, Ely and Geall 2015).

In the next section, I introduce two perspectives on innovation and change which originate from the field of Science and Technology Studies (STS). These perspectives may help overcoming some of the weaknesses of the innovation systems approach by critically asking how change can occur within a system, and how stability is maintained: how things, people and concepts are assembled, mobilised and circulated into what they are known as today.

Science and technology studies perspectives on innovation

One part of the endeavour of this book is to gain a better understanding of how innovation in China's wind and solar industries can be understood and conceptualised. Innovation systems thinking tends to see the world through fixed boxes which by and large remain the same over time. This is why it can be said to reside within a structuralist or institutionalist ontology. The basic assumption within science and technology studies (STS) is that science and technology are social activities, embedded in society like any other activities (Sismondo 2004). This means that science and knowledge are not 'truths' deducted from nature, but are really influenced, for instance, by the community at large, politics or rhetorical work which aim to convince people (Sismondo 2004). Nevertheless, STS is not a single well-defined body of work. According to Russell and Williams (2002, p. 61) one broad strand within STS has sought ways of 'reconciling and articulating different levels of analysis' whilst another, by contrast, has 'retained a strong scepticism towards the very idea [of structures] (...)'. Here, I highlight two perspectives which focus on innovation, namely sustainability transition perspectives and actor-network theory (ANT). Sustainability transition perspectives can be said to belong in the first, reconciling strand, whilst ANT has remained sceptical to the notion of levels and structures, and can be said to have a 'flat' ontology, meaning that there are no levels. To summarise, innovation system approaches believe in institutions and structures, sustainability transitions has sought a 'middle way' and ANT is 'flat'. The difference between these three perspectives allows for an interesting discussion of how innovation can be understood and conceptualised in China.

Sustainability transition perspectives

Sustainability transition studies focus explicitly on technological change and sustainability. Unlike technological transitions-as-usual, which, by and large, are emergent, sustainability transitions are intentional and involve the coordination of several actors aiming at an agreed-upon purpose (Markard, Raven and Truffer 2012). The single most influential concept in transition research is the 'sociotechnical regime' (Markard, Raven and Truffer 2012), which builds heavily on Nelson and Winter's (1982) concept of technological regimes (Geels 2002). This implies that transition studies have much in common with innovation systems thinking. In general, technological regimes are taken to encompass all of the factors which influence, support and constrain the development of a technology, including production and thought processes, routines, knowledge and learning (Dosi and Nelson 1994). Conceptualisations of technological regimes are frequently referred to in descriptions of the way in which learning and the knowledge environment shape the direction in which technology develops through tacit agreements of involved agents (Malerba 2005).

When looking at energy industries, insights on regimes become evident since large infrastructures are already in place, and these do not necessarily cater for intermittent renewable energy technologies (Kemp 1994). The renewal of energy technologies happens especially slowly because capital expenses are large and investments are long-term, for example, ranging from 30 to 40 years for an average coal-fired power station (Jørgensen and Münster 2010, p. 10). Most countries' electric power systems are fuelled by fossil fuels and have developed over time, creating vested benefits for the continued use of fossil fuels. This development has been termed a 'carbon lock-in' and is considered one of the prime reasons why new renewable energy technologies face difficulty (Unruh 2000). Such concerns served as the backdrop to how thinking on sustainability transition perspectives emerged (Kemp 1994).

Two central approaches within the 'sustainability transitions' approach are the 'strategic niche management' and 'transitions management' approaches, which appear to have been inspired directly by the concern which the current technological situation is environmentally unsustainable (Kemp 1994; Kemp and Rotmans 2005). The strategic niche management framework is based on historical studies of technological regime shifts (Kemp, Schot and Hoogma 1998). From this, Kemp, Schot and Hoogma (1998, p. 183) found that there are two important lessons: first, 'entrepreneurs/system builders and niches play an important role in the transition process[es]'; and second, there is an 'availability of niches or domains for application'. This latter point means that there needs to be a market for new technologies so that a 'protected space' can enable the niche to develop. This was, for instance, the case with many important technologies of the twentieth century which received special support through military or other state funding, such as radios, aircrafts and computers (Kemp, Schot and Hoogma 1998; Mazzucato 2013). Kemp, Schot and Hoogma (1998, p. 186) therefore proposed the following definition:

> strategic niche management is the creation, development and controlled phase-out of protected spaces for the development and use of promising technologies by means of experimentation, with the aim of (1) learning about the desirability of the new technology and (2) enhancing the further development and the rate of application of the new technology.

The definition points out that a protected space must eventually be phased out, since the technology will either become successful and no longer need protection, or it will fail, in which case further support would not make sense. Three elements characterise successful niche emergence and growth: visions and expectations, networks and learning (Seyfang and Longhurst 2013). Visions and expectations must be shared, networks must encompass several stakeholders and learning 'should contribute not only to everyday knowledge and expertise, but also to "second-order learning" wherein people question the assumptions and constraints of mainstream systems altogether' (Seyfang and Longhurst 2013, p. 882). The role of the government as a supporter, protector and nourisher of niches is relatively implicit, as there are few other actors with the means, foresightedness and courage to take on such a role, as pointed out by Mazzucato (2013). According to this approach, we can expect the Chinese government to take an active role in the formation of potential wind and solar niches.

Transition management was originally formulated as a management strategy for public decision-makers and can be said to be an action- or process-oriented version of the MLP (Rotmans, Kemp and van Asselt 2001). It places itself between bottom-up approaches—such as strategic nice management—and top-down approaches which entail comprehensive planning and political steering (Kemp, Loorbach and Rotmans 2007). Overarching policy steering is identified as difficult due to challenges with distributed control (actors have different levels of authority, depending on the area), dissent (actors have diverging opinions about sustainability) and political myopia (politicians tend to be short-sighted, but transitions take time) (Kemp, Loorbach and Rotmans 2007). By combining elements of long-term planning with incrementalism, the transition management approach sets out to

> create a societal movement through new coalitions, partnerships and networks around arenas that allow for building up continuous pressure on the political and market arena to safeguard the long-term orientation and goals of the transition process.
>
> (Loorbach and Rotmans 2010, p. 239)

Transition management approaches also place significant emphasis on the role of guiding visions, which can motivate and coordinate actors (Smith, Stirling and Berkhout 2005). Briefly put, the transition management approach is concerned with problems which relate to both short-term concerns and long-term imperatives, and it is explicitly pointed out that the 'government can and should assume a leading role in transition management' (Rotmans, Kemp and van Asselt 2001, p. 25). The role of government should be that of reflexive and evolutionary management, with government actors taking on multiple roles, such as facilitators, stimulators, controllers or directors depending on the perceived stage of transition (Markard, Raven and Truffer 2012; Rotmans, Kemp and van Asselt 2001).

The transition management perspective is closely related to the term 'reflexive governance' (Voß and Bornemann 2011). Reflexive governance means that governments understand themselves to be part of the dynamics which they govern. As pointed out by Voß and Kemp (2005, p. 4) reflexive governance incorporates feedback

> by opening problem-handling processes for diverse knowledge, values and resources of influence in order to learn about appropriate problem-definitions, targets and strategies of governance for sustainable development. As such reflexive governance is about the organisation (modulation) of recursive feedback relations between distributed steering activities.

Hence, reflexive governance means to be open for learning and changing policy according to what the (new) aim is, and what proves to be effective policy measures. Indeed, reflexive governance appears to be a way of learning so that policy can be optimised according to the sustainability criteria at hand. Therefore, experimentation and learning by doing are approaches which transition management scholars emphasise (Voß and Bornemann 2011; Voß and Kemp 2005).

Both of these 'management approaches' draw on the importance of the future in the shape of plans or visions which would define suitable niches. Expectations and visions not only formulate a desired future, but also constitute and accommodate it (Borup *et al.* 2006). Expectations at once legitimise a technology, indicate a development direction and help coordinate interests which can materialise into technical and scientific activities and products (van Lente 2012). Van Lente (2000) describes a dynamic of 'promise and requirement', implying that commitments become part of a shared agenda which requires action in itself. When promises are made, they subsequently become required achievements and, ultimately, necessary to complete or to continue working on the achievement; in other words, they become self-fulfilling prophecies (van Lente 2000). This is not to say that any vision may become self-fulfilling on its own, but rather, there is often substantial and active work behind making the future and aligning expectations.

Although many visions fail to materialise (see, e.g., Geels and Smit 2000), many visions do succeed. What characterises a successful vision, and how is it performed? Several researchers have addressed these questions. Berkhout (2006) describes expectations as bids for the future which offer a potentiality requiring endorsement from others to be actualised. In this manner, interpretive flexibility persists, increasing the likelihood of greater support and potentially becoming larger and more normative and collective visions located in 'art and literature, public and political discourses, statements and appeals from business, civil society and government' (Berkhout 2006, p. 307). Skjølsvold (2014) stresses that futures are resources which spokespeople of a technology can draw on to reach goals. These futures are performative in two ways: They enrol and convince new actors, and they gradually transform the idea of what a technology could mean for the future and nest it with more purposes than were inherent to the initial idea. In this sense, a wind turbine or a PV panel is not only 'electricity'; it is also a 'sustainable lifestyle', a 'green' image which China can show to the world.

Expectations may also be contradictory and may fight to conquer the future to stay relevant (Bakker, van Lente and Meeus 2011; Eames *et al.*, 2006), which is not necessarily a weakness, as scenarios are malleable and can be used to build support from various quarters (Sovacool and Ramana 2015). Moreover, as has been noted by several expectations' scholars (e.g., Geels and Smit 2000; Gjøen 2001), some promises are set unrealistically high to generate interest and a protected 'space' for a technology. Though unrealistic, these promises are useful for the spokespeople of a technology, as they may impact the way people think about a topic (Berkhout 2006; Pollock and Williams 2010). As Geels and Smit (2000, p. 883) put it, 'some future speculations do not strive for truth or accuracy, but are meant to influence specific social processes in technological developments'. These expectations are therefore part of *strategic games*, wherein the spokespeople risk that their 'bluff' will be called.

As expectations become taken for granted, they become able to mobilise a larger number of actors—even actors who do not necessarily benefit directly from the expectation (Konrad 2006). Eames *et al.* (2006) use the word 'vision' to denote a coherent image of the future intended to generate long-term action, whereas expectations of such images are less formal and include more fragmented beliefs about the future. The authors find that the vision of a 'hydrogen economy' is underpinned by a range of narratives such as those of democracy, independence, environment and power. As the vision is brought down to a local level, however, each narrative is drawn upon to generate interest or disinterest, impacting the way in which the guiding vision is perceived. Local expectations, therefore, impact the guiding vision. In this book, I use the concept 'sociotechnical imaginary' to denote these larger 'visions' and 'collective images', whilst 'expectations' pertain to more local, and industry-specific processes. Jasanoff and Kim (2009 p. 120) define sociotechnical imaginaries as 'collectively imagined forms of social life and social order reflected in the design and fulfilment of nation-specific scientific and/or technological projects'. Jasanoff and Kim (2009) emphasise the importance of governments in simultaneously describing and prescribing an attainable and desirable future. Several recent studies have highlighted the different roles of the government in establishing collective images. For instance, Alkemade and Suurs (2012) argue that governments are more prominent in discourses and expectations related to sustainable technologies. Bakker, van Lente and Meeus (2011, p. 159) describe an 'arena of expectations' wherein expectations are voiced and tested and spokespeople and potential receivers of a technology are 'confronted with experience, knowledge, and interests'. Within these 'arenas of expectations', state-industry arrangements function to make certain expectations more credible than others through exchanges between the government, industry and consultancies (Levidow, Borda-Rodriguez and Papaioannou 2014).

The 'sociotechnical imaginaries' concept emerged from a growing recognition that expectations of future possibilities are embedded in the organisation and practice of science and technology (Jasanoff and Kim 2009, p. 122). Imaginaries are characterised as less instrumental than policy agendas, but more instrumental than grand narratives, and they 'reside in the reservoir of norms and discourses, metaphors and cultural meanings out of which actors build their policy preferences' (Jasanoff and Kim 2009, p. 123). Imaginaries are futuristic and direct actors towards a desirable future. State power is a central 'director' of desirable futures through various means, such as the 'selection of development priorities, the allocation of funds, the investment in material infrastructures, and the acceptance or suppression of political dissent' (Jasanoff and Kim 2009, p. 123). Moreover, 'the resulting politics of science and technology may shape not only the narrow issues surrounding those specific enterprises but also wider social and political understandings about a nation's past, present and future' (Jasanoff and Kim 2009, p. 124). Thus, sociotechnical imaginaries involve direct state intervention and come into play when forms of nationhood are imagined together with a technological project, such as the development of wind and solar power, as is the case in this book.

Both the strategic niche management and transitions management highlight the role of the government in innovation and transition processes. Lately, the important role of the government as a facilitator of innovation has been emphasised by scholars such as Mazzucato (2013). In China, the government is a key actor, and sustainability transition thinking represents an interesting model of how to think about the role of the government in transition processes. Nevertheless, the sustainability transitions framework has also been critiqued from several holds, particularly in asserting that a transition is something which indeed can be managed, when knowledge about exactly where we want to transition to and how to get there is uncertain (e.g., Shove and Walker 2007). This is also discussed by Labussière and Nadaï (2018, p. 15), who argue that the 'transition management' approaches are problematic 'because the levels and the dynamics to be described are partly defined beforehand'. Instead, transition is seen as a 'process' rather than as a 'trajectory'. Thus, they argue that a relational approach that avoids predefining levels of analysis, but rather follows the making of transition issues, and their 'emergence as political objects' is to prefer (Labussière and Nadaï 2018, p. 16). Avoiding predefined 'boxes' for analysis is perhaps also particularly important in the context of China. Transition in the sense of transitioning from communism to capitalism, for instance, is easily thought of as an ideological project—with the aim of integrating China into a kind of Western core. Thus, transition thinking could also be mistaken for a political project of 'rendering European', something that the CCP clearly is not interested in. For these reasons, this book draws on a perspective within STS called actor-network theory, which is relational and 'flat' meaning that it does not have predefined levels of analysis.

Actor-network theory

> Innovation is the art of interesting an increasing number of allies who will make you stronger and stronger.
>
> (Akrich, Callon and Latour [2002a, p. 205])

Actor-network theory (ANT) takes as a point of departure that technology does not develop in a vacuum, nor is one single person responsible for inventing and developing it. By closely examining and 'opening up' the innovation process, ANT analyses the way in which technologies are compiled to become innovations. When scientific facts and technologies are presented to the world, they appear fixed and nonnegotiable. However, in the process of becoming stabilised as facts and artefacts, they go through a period of contestation and controversy. For ANT, controversies are sites in which social connections are traceable; therefore, they are convenient for understanding the way in which innovations are products of negotiations or translations (Latour 2005). ANT is concerned with the study of change, and the way in which something becomes 'stabilised' into its current form. As pointed out by Star (1991, p. 38), the advantage of actor-network theory is to move from the experience which an enormous amount of resources are massed in order to build technologies, 'to the fact that "it might have been otherwise"—there is nothing necessary or inevitable about any such science or technology, all constructions are historically contingent, no matter how stabilized'. ANT highlights the alliances behind technical choice and, within these alliances, the exertion of agency. As Callon (1986, p. 216) notes: 'To speak for others is to first silence those in whose name we speak'. For this reason, ANT points out that there are always 'spokespersons' of a technology who engage in an arduous process of translation.

Translation is a central concept in ANT, and, simply put, it pertains to marshalling people and resources to back up a claim so it becomes fact (Latour 1987). Callon (1986) distinguishes four moments of the translation process: problematisation, interessement, enrolment and mobilisation. Problematisation means framing an issue so that the actor who is framing becomes an 'obligatory passage point' to the interested people. However, people might not know they are interested, and this is the point of interessement: the roles and identities of the different actors must be defined so they see the need of this particular idea, product, innovation or similar entity (in this case, solar and wind energy). Enrolment is the result of successful interessement. As actors negotiate the necessity of the entity at hand, they must convince and test the roles of each involved party so that agreement and conformity with the initial 'problematisation' are achieved. Mobilisation includes more people and things into the actors' undertaking and involves a more widespread acceptance of the entity outside the initial 'circle' of involved actors (Callon 1986). These four 'moments' are neither consecutive nor simultaneous; they appear throughout the translation process, and each 'moment' may be drawn on at any point in time. Displacement is a central concept in translation, because at all stages in the process, actors, goals, interests and objects are displaced in order to conform to the initial problematisation (Latour 1983). This displacement is necessary to promote the role of the actor, or spokesperson, of the technology.

A common understanding of innovation is that it is an invention which is taken into use. With this definition, innovation only exists if it is 'stabilised' and accepted by people. By better understanding the practices which innovators are involved in, we may say something about the reason some inventions get stabilised whilst others do not. This is a recurring topic in the work of Latour (e.g., 1983; 1987), as he frequently refers to skilled and cunning innovators such as Pasteur, Diesel or Edison, and the way in which they were able to mobilise resources and people to support their ideas. Let us consider the example of 3M's Post-it notes to illustrate this point. Spence Silver was a 3M scientist who, through experimentation, discovered the weak glue which eventually became the innovative Post-it notes (Garud and Karnøe 2001). Working in a company which produced 'glue that sticks', Silver was relatively unclear what the weak glue would be useful for, but he was sure he had 'a solution looking for a problem' (Garud and Karnøe 2001, p. 13). Silver met resistance or indifference everywhere he went with his glue, but he did not give up trying to find a problem which would complement his solution—problematisation. With time, Silver managed to convince one fellow scientist of the glue's usefulness. Ten years after the discovery, a third scientist, Art Fry, also became interested (Garud and Karnøe 2001)—interessement. Together, they explored different uses of the glue, such as putting weak glue on an entire board so that paper notes could stick to it, but their 3M colleagues were not impressed.

It was not until one day, after having mixed up his pages at a choir practice (so the story goes), that Fry came up with the idea of putting a bit of glue on small sheets of paper, and the idea of Post-it notes as we know them was shaped (Garud and Karnøe 2001)—enrolment. From this point, the notes gained momentum and the three scientists were able to mobilise more people. Failing to convince their marketing colleagues, they decided to give away free samples of the small yellow notes to all of the personal 3M secretaries and even competing firms (Akrich, Callon and Latour 2002b). Soon, the secretaries started asking for more, and this finally caught the interest of 3M's marketing division and, eventually, the company directors agreed to give the product a go (Garud and Karnøe 2001)—mobilisation. Finally, in 1976, 12 years after the glue was discovered, the Post-it note was mass-manufactured (Garud and Karnøe 2001).

This short example shows how 'failed' glue was transformed through a series of displacements; 'it is not failed' to 'it might be used' to 'it is useful'. Innovation can be interpreted as this process of translation and displacement which renders something useful to others by aligning interests and goals (Latour 1987). In this process, successful enrolment of both humans and non-humans was a criterion for success: had the glue turned out to be poisonous, had Silver been fired, had the glue board idea been a success, had Fry not been an avid choir singer and so on, the Post-it note might not have been created. What we see in this case is a co-production of the object and its environment through a slow process of linking glue, yellow paper, users, distributors and procedures. Instead of inventions being selected by their environment, as evolutionary innovation perspectives claim (Nelson and Winter 1982), inventions and the environment are co-produced: 'the environment is produced at the same time as the innovation that it is going to judge' (Akrich, Callon and Latour 2002b, p. 211).

This co-production requires the spokespersons of the technology to be convincing and persistent. Such spokespersons cannot be perturbed by the obstacles they face along the way; they must turn resistance into new opportunities and new allies so that circumstances are no longer unfavourable. For this reason, Akrich, Callon and Latour (2002a, p. 218) say that 'the innovation which succeeds is the one which comes to master it by choosing good speakers'. These speakers, in turn, delegate agency to intermediaries, such as the weak glue which Silver, the spokesperson, believed in and trusted. Nevertheless, there is always uncertainty connected with intermediaries, and the innovator can 'never completely know who or what they are representing, and whether indeed they (intermediaries) really are representing them'

(Akrich, Callon and Latour 2002a, p. 222). Problematisation, interessement, enrolment and mobilisation are highly uncertain processes which require skill and time in order to successfully create a path for a new technology. An innovation, in this context, is co-produced with its environment, rather than seen as an outcome of specific types of institutions or knowledge. Innovation as co-production is a central tenet which will inform the understanding of innovation in this book.

Latour (2005) introduces the central concepts of 'mediators' and 'intermediaries' to explain ANT. An intermediary is defined as 'what transports meaning or force without transformation—defining its inputs is enough to define its outputs' (Latour 2005, p. 39). In other words, an intermediary may be taken for what it is. Mediators, on the other hand, 'transform, translate, distort, and modify the meaning or the elements they are supposed to carry'. Mediators are therefore more interpretive and do not carry a fixed meaning which is easily and faithfully transported. What distinguishes ANT from other social theories, Latour claims, is that ANT does not take social aggregates as intermediaries—signposts which carry meaning—but as mediators which, only as rare exceptions, are understood as intermediaries. Mediators may be interpreted here as actors, whilst intermediaries can be 'anything passing between actors which defines the relationship between them' (Callon 1991, p. 134).

The twist is that mediators and intermediaries may also be hybrids, meaning that they may in one situation appear as an intermediary and later appear as a mediator, and vice versa. For instance, 'transition' may be regarded as a concept intended to coordinate likeminded actors towards a similar direction (i.e., an intermediary), but 'transition' may also be used as a means to achieve a particular, and unrelated aim, i.e., staying in power or increasing influence (i.e., a mediator). The task at hand is therefore to understand whether entities are behaving as mediators or intermediaries, and this is a matter of empirical investigation (Callon 1991). In other words, intermediaries are representatives which render displacements easier, and they create bridges, or equivalences, between two previously unconnected items which result in the designation of spokespeople (Callon 1986). These representatives may be things, numbers and figures—for example statistics showing that fishermen are suffering from decreasing yield and are thus in need of scallop farming. Technologies, things and nonhumans can therefore be taken to be masses—facts and artefacts—that help hold an actor-network together.

Intermediaries may also be specialised agencies which act in the interest of technology spokespeople, such as experts, consultants or advisory agencies. In this latter case, the distinction between spokespeople and intermediaries is more blurred, as intermediaries also act in self-interest and become spokespeople of the technology when their business model is based on support of a specific technology. If we take this as a point of departure to understand technology transfer, for instance, technology 'adoption' can be understood to be synonymous with adaptation, because its need and use must be constantly negotiated between the spokespeople of a technology and the surrounding community in order for it to spread (Akrich, Callon and Latour 2002a; Akrich, Callon and Latour 2002b). Following this logic, it makes more sense to talk about technology *translation* rather than technology *transfer*. In the case of China's wind and solar industries, this insight may be interpreted as an attempt by technology spokespeople to push procedures and practices on China. In other words, China would not be 'imitating' Western firms, but rather accepting some aspects of a technology and negotiating other aspects and ending up with a locally negotiated hybrid as a result. Such a process by necessity also involves certain risks.

There are also other ways to show that a lot of work is needed for a technology to come into existence through actor-networks. Latour (2004) draws a distinction between matters of fact (or objects/technologies) and matters of concern (or *Things*). A Thing is seen as a gathering—an arena for contestation, negotiation and agreement—whilst an object is what is 'out there, unconcerned by any sort of parliament, forum, agora, congress, court' (Latour 2004, p. 236). This distinction shows that scientific facts and technologies only exist in their present form because they were first matters of concern. As Latour (2004, p. 247) puts it: 'All objects are born things, all matters of fact require, in order to exist, a bewildering variety of matters of concern'. Rather than accepting matters as facts, we should look at them as matters of concern which are open and inclusive arenas in which participants (humans and nonhumans) can gather, instead of limited and black-boxed objects. ANT, in this light, can be seen as a process of detecting 'how many participants are gathered in a thing to make it exist and to maintain its existence' (Latour 2004, p. 246). For this reason, Latour (2005) talks about assemblages, wherein new associations are made between humans, nonhumans, knowledge, practices and organisations, and technologies emerge as stabilised entities.

Assemblages require a lot of work by all involved human and nonhuman actors in order to be linked together. This type of work happens continually, on an ad hoc basis, and it is performative, that is, it 'shape[s] actors and actors' relations as well as their practices and understandings' (Hojem and Lagesen 2011, p. 124). In other words, assemblage may be understood as a process of 'recursive self-assembling in which the elements put together are not fixed in shape, do not belong to a larger pre-given list but are constructed at least in part as they are entangled together' (Law 2004, p. 42). Therefore, a central argument for Latour (2005, p. 247) is that the social is 'but a moment in the long history of assemblages'. However, we do not deliberately and knowingly work to uphold an assemblage. We are always, as Latour points out elsewhere, 'slightly surprised' by our actions (Latour 1999, p. 281). We should therefore understand assemblage as a tentative and hesitant unfolding, which is 'at most only very partially under any form of deliberate control' (Law 2004, p. 42).

If we take these insights from ANT on how constellations are brought and held together by human and nonhuman actors and the associations between them, we arrive at an understanding of knowledge and learning which is markedly different from the one introduced by innovation system scholars. If knowledge is co-produced by the actors and their actor-worlds, then learning can be understood as an enactment of knowledge, or knowing-as-displacing (Law 2000). This type of learning can be understood as the enactment of different types of knowledge, in different locations. As Law (2000, p. 354) puts it, an ANT understanding of knowing-as-displacing 'recognizes that its subjects and objects are no longer fixed but are better understood as shifting and elusive processes'. In other words, learning can be understood as a heterogeneous process which is part of a mutual inclusion with other enactments and that is characterised by uncertainty. With such an understanding, learning is relative: what type of knowledge is necessary where, for which purpose and to whose reality does this knowledge belong? Knowledge can in this perspective not be viewed as an ideal type to strive for but must in each case be studied empirically by understanding the strategies and actions of the actors in question.

Star (1991, p. 52) points out that: 'Power is about whose metaphor brings worlds together, and holds them there'. By using ANT, the task at hand for the cases in this book is to trace the various ways in which something is accepted or taken for granted, such as standards, procedures and practices, and the various ways in which what is taken for granted may be a product of the 'spokespersons' behind a technology. The task is, in other words, to try to uncover whose metaphors are important, and why. What negotiations take place between new users and old providers? What translation strategies are used?

We could say that the main difference between ANT and the sustainability transitions perspectives is the fundamental perspective on how change occurs (or not occurs): ANT focuses on how people and things constantly and recurringly work to assemble a fact or a concept which may become stable. Hence, the focus is often on processes produced through controversy which may or may not lead to stabilisation. In sustainability transition perspectives things tend to remain the same, within sociotechnical regimes, and only a massive effort can induce change. Thus, the focus is often on how a sociotechnical regime is overtaken or changed because of forces from the 'outside'. To summarise, the ANT approach can capture controversies which may lead to innovation, but controversies may continue because the parties involved still negotiating what that 'innovation' means. Sustainability transition approaches are more static but better at seeing the larger picture and linkages between domains which may facilitate or complicate (an already given understanding of) innovation.

References

Abramowitz, M., 1986. Catching up, forging ahead, and falling behind. *The Journal of Economic History*, 46(2), pp. 385–406.

Akrich, M., Callon, M. and Latour, B., 2002a. The key to success in innovation part i: the art of interessement. *International Journal of Innovation Management*, 6(2), pp. 187–206.

Akrich, M., Callon, M. and Latour, B., 2002b. The key to success in innovation part ii: the art of choosing good spokespersons. *International Journal of Innovation Management*, 6(2), pp. 207–225.

Alkemade, F. and Suurs, R.A.A., 2012. Patterns of expectations for emerging sustainable technologies, *Technological Forecasting & Social Change*, 79, pp. 448–456.

Amsden, A.H., 1989. *Asias's Next Giant. South Korea and Late Industrialization*. New York & Oxford: Oxford University Press.

Bakker, S., van Lente, H. and Meeus, M.T.H., 2011. Arenas of expectations for hydrogen technologies. *Technological Forecasting & Social Change*, 78, 152 and 162.

Berkhout, F., 2006. Normative expectations in systems innovation. *Technology Analysis & Strategic Management*, 18, pp. 299–311.

Bijker, W.E. and Law, J. (eds.), 1992. *Shaping Technology/Building Society: Studies in Sociotechnical Change*. Cambridge, MA: MIT Press.

Borup, M., Brown, N., Konrad, K. and Van Lente, H., 2006. The sociology of expectations in science and technology. *Technology Analysis & Strategic Management*, 18(3–4), pp. 285–298.

Bruland, K. and Mowery, D.C., 2005. Chapter 13. Innovation through time. In J. Fagerberg, D.C. Mowery and R.R. Nelson (eds) *The Oxford Handbook of Innovation*. Oxford: Oxford University Press, pp. 349–379.

Callon, M., 1986. Some elements of a sociology of translation: domestication of the scallops and the fishermen of St. Brieuc Bay. In J. Law (ed.) *Power, Action and Belief: A New Sociology of Knowledge?* London: Routledge, pp. 196–223.

Callon, M., 1991. Techno-economic networks and irreversibility. In J. Law (ed.) *A Sociology of Monsters: Essays on Power, Technology and Domination*. London and New York: Routledge, pp. 132–165.

Chang, H.J., 2002. *Kicking away the Ladder: Development Strategy in Historical Perspective*. London: Anthem Press.

Chang, H.J., 2007. Globalization, global standards, and the future of East Asia. In J.-S. Shin (ed.) *Global Challenges and Local Responses. The East-Asian Experience*. London: Routledge.

Cohen, W.M. and Levinthal, D.A., 1990. Absorptive capacity: a new perspective on learning and innovation. *Administrative Science Quarterly*, 35(1), pp. 128–152.

Dittmer, L. and Liu, G. (eds), 2006. *China's Deep Reform: Domestic Politics in Transition*. Lanham, Boulder, New York, Toronto, Oxford: Rowman & Littlefield.

Dodgson, M., 1993. Organizational learning: a review of some literatures. *Organization Studies*, 14(3), pp. 375–394.

Dosi, G. and Nelson, R.R., 1994. An introduction to evolutionary theories in economics. *Journal of Evolutionary Economics*, 4(3), pp. 153–172.

Eames, M., Mcdowall, W., Hodson, M. and Marvin, S., 2006. Negotiating contested visions and place-specific expectations of the hydrogen economy. *Technology Analysis & Strategic Management*, 18(3–4), pp. 361–374.

Edquist, C., 2005. Chapter 7. Systems of innovation: perspectives and challenges. In J. Fagerberg, D.C. Mowery, and R.R. Nelson (eds) *The Oxford Handbook of Innovation*. Oxford: Oxford University Press.

Fagerberg, J. and Godinho, M.M., 2005. Chapter 19. Innovation and catching-up. In J. Fagerberg, D.C. Mowery, and R.R. Nelson (eds) *The Oxford Handbook of Innovation*. Oxford: Oxford University Press, pp. 514–542.

Fagerberg, J., Mowery, D.C. and Nelson, R.R., 2005. *The Oxford Handbook of Innovation*. Oxford: Oxford University Press.

Garud, R. and Karnøe, P., 2001. Path creation as a process of mindful deviation. In R. Garud and P. Karnøe (eds) *Path Dependence and Creation*. Mahwah: Lawrence Erlbaum Associates, Inc.

Garud, R. and Karnøe, P., 2003. Bricolage versus breakthrough: distributed and embedded agency in technology entrepreneurship. *Research Policy*, 32(2), pp. 277–300.

Geels, F.W., 2002. Technological transitions as evolutionary reconfiguration processes: a multi-level perspective and a case-study. *Research Policy*, 31(8–9), pp. 1257–1274.

Geels, F.W. and Smit, W.A., 2000. Failed technology futures: pitfalls and lessons from a historical survey, *Futures*, 32, pp. 867–885.

Gjøen, H., 2001. Gasstanker. En sosiologisk studie av visjoner og virkelighetskonstruksjoner knyttet til bruk av naturgass i Norge, Norwegian University of Science and Technology (NTNU).

Goody, J., 1996. *The East in the West*. Cambridge: Cambridge University Press.

Hojem, T.S.M. and Lagesen, V.A., 2011. Doing environmental concerns in consulting engineering. *Engineering Studies*, 3(2), pp. 123–143.

Hussain, A., 2002. Demographic transition in China and its implications. *World Development*, 30(10), pp. 1823–1834.

Isaksen, A. and Nilsson, M., 2012. Combined innovation policy: linking scientific and practical knowledge in innovation systems. *European Planning Studies*, 21(12), pp. 1919–1936.

Jasanoff, S. and Kim, S.H., 2009. Containing the atom: sociotechnical imaginaries and nuclear power in the United States and South Korea. *Minerva*, 47(2), pp. 119–146.

Jensen, M.B., Johnson, B., Lorenz, E. and Lundvall, B.Å., 2007. Forms of knowledge and modes of innovation. *Research Policy*, 36(5), pp. 680–693.

Johnson, B., 2010. Chapter 2: institutional learning. In B.Å. Lundvall (ed.) *National Systems of Innovation: Towards a Theory of Innovation and Interactive Learning*. London: Anthem Press.

Jørgensen, B.H. and Münster, M., 2010. *Nye energiteknologier. Forskning, udvikling og demonstration*. Danmarks Tekniske Universitet (DTU), Risø Nationallaboratoriet for Bæredygtig Energi. Available at: www.bibliotek.dtu.dk/Inst/MAN/Service/Telefonbog.aspx?lg=showcommon&id=270534.

Kamp, L.M., Smits, R.E.H.M. and Andriesse, C.D., 2004. Notions on learning applied to wind turbine development in the Netherlands and Denmark. *Energy Policy*, 32(14), pp. 1625–1637.

Kemp, R., 1994. Technology and the transition to environmental sustainability: the problem of technological regime shifts. *Futures*, 26(10), pp. 1023–1046.

Kemp, R., Loorbach, D. and Rotmans, J., 2007. Transition management as a model for managing processes of co-evolution towards sustainable development. *International Journal of Sustainable Development & World Ecology*, 14(1), pp. 78–91.

Kemp, R. and Rotmans, J., 2005. The management of the co-evolution of technical, environmental and social systems. In M. Weber and J. Hemmelskamp (eds) *Towards Environmental Innovation Systems*. Berlin: Springer Berlin Heidelberg, pp. 33–55.

Kemp, R., Schot, J. and Hoogma, R., 1998. Regime shifts to sustainability through processes of niche formation: the approach of strategic niche management. *Technology Analysis & Strategic Management*, 10(2), pp. 175–195.

Kim, L., 1997. *Imitation to Innovation: The Dynamics of Korea's Technological Learning*. Boston: Harvard Business School Press.

Kline, S.J. and Rosenberg, N., 1986. An overview of innovation. In R. Landau and N. Rosenberg (eds) *The Positive Sum Strategy: Harnessing Technology for Economic Growth*. Washington, D.C.: The National Academies Press.

Konrad, K., 2006. The social dynamics of expectations: the interaction of collective and actor-specific expectations on electronic commerce and interactive television. *Technology Analysis & Strategic Management*, 18(3–4), pp. 429–444.

Labussière, O. and Nadaï, A. (eds), 2018. *Energy Transitions: A Socio-Technical Inquiry*. Energy, Climate and the Environment Series. Palgrave Macmillan: Cham.

Latour, B., 1983. Give me a laboratory and i will raise the world. In K.D. Knorr-Cetina and M. Mulkay (eds) *Science Observed. Perspectives on the Social Studies of Science*. London, Beverly Hills, New Delhi: Sage Publications, pp. 141–170.

Latour, B., 1987. *Science in Action*. Cambridge, MA: Harvard University Press.

Latour, B., 1999. *Pandora's Hope: Essays on the Reality of Science Studies*. Cambridge, MA: Harvard University Press.

Latour, B., 2004. Why has critique run out of steam? From matters of fact to matters of concern. *Critical Inquiry*, 30(2), pp. 225–248.

Latour, B., 2005. *Reassembling the Social. An Introduction to Actor-Network-Theory*, Oxford, New York: Oxford University Press.

Law, J., 2000. Comment on Suchman, and Gherardi and Nicolini: knowing as displacing. *Organization*, 7(2), pp. 349–354.

Law, J. 2004. *After Method. Mess in Social Science Research*. London: Routledge.

Levidow, L., Borda-Rodriguez, A. and Papaioannou, T., 2014. UK bioenergy innovation priorities: making expectations credible in state-industry arenas. *Technological Forecasting and Social Change*, 87, pp. 191–204.

Liu, X. and White, S., 2001. Comparing innovation systems: a framework and application to China's transitional context. *Research Policy*, 30, pp. 1091–1114.

Loorbach, D. and Rotmans, J., 2010. The practice of transition management: examples and lessons from four distinct cases. *Futures*, 42(3), pp. 237–246.

Lundvall, B.Å., 1988. Innovation as an interactive process: from user-producer interaction to the national system of innovation. In G. Dosi *et al.* (eds) *Technical Change and Economic Theory*. London and New York: Pinter Publishers, pp. 349–370.

Lundvall, B.Å., 2007. National innovation systems—analytical concept and development tool. *Industry & Innovation*, 14(1), pp. 95–119.

Lundvall, B.Å. (ed.), 2010. *National Systems of Innovation: Towards A Theory of Innovation and Interactive Learning*. The Anthem Other Canon series. London: Anthem.

Lundvall, B.Å., 2011. Notes on innovation systems and economic development. *Innovation and Development*, 1(1), pp. 25–38.

Lundvall, B.Å., Joseph, K.J., Chaminade, C. and Vang, J. (eds), 2009. *Handbook of Innovation Systems and Developing Countries: Building Domestic Capabilities in a Global Setting*. Cheltenham: Edward Elgar Publishing.

Malerba, F., 2005. Chapter 14: sectoral systems: how and why innovation differs across sectors. In J. Fagerberg, D.C. Mowery and R.R. Nelson (eds) *The Oxford Handbook of Innovation*. Oxford: Oxford University Press.

Markard, J., Raven, R. and Truffer, B., 2012. Sustainability transitions: an emerging field of research and its prospects. *Research Policy*, 41, 955–967.

Mazzucato, M., 2013. *The Entrepreneurial State: Debunking Public vs. Private Myths in Risk and Innovation*. London, New York, Delhi: Kindle Edition.

Naughton, B., 2007. *The Chinese Economy. Transitions and Growth*. Cambridge, Mass.: MIT Press.

Nelson, R.R., 2004a. The challenge of building an effective innovation system for catch-up. *Oxford Development Studies*, 32(3), pp. 365–374.

Nelson, R.R., 2004b. The market economy, and the scientific commons. *Research Policy*, 33(3), pp. 455–471.

Nelson, R.R. and Winter, S.G., 1982. *An Evolutionary Theory of Economic Change*. Cambridge, MA: Belknap Press.

Perez, C., 2010. Technological revolutions and techno-economic paradigms. *Cambridge Journal of Economics*, 34(1), pp. 185–202.

Perez, C., 2013. Innovation systems and policy for development in a changing world. In J. Fagerberg, B.R. Martin and E.S. Andersen (eds) *Innovation Studies: Evolution and Future Challenges*. Oxford: Oxford University Press, pp. 90–111.

Perez, C. and Soete, L., 1988. Catching up in technology: entry barriers and windows of opportunity. In G. Dosi, C. Freeman, R.R. Nelson, G. Silverberg and L. Soete (eds) *Technical Change and Economic Theory*. London and New York: Pinter Publishers, pp. 458–479.

Pollock, N. and Williams, R., 2010. The business of expectations: how promissory organizations shape technology and innovation. *Social Studies of Science* 40, pp. 525–548.

Rosendal, G.K., 2001. Impacts of overlapping international regimes: the case of biodiversity. *Global Governance*, 7(1), pp. 95–117.

Rotmans, J., Kemp, R. and van Asselt, M., 2001. More evolution than revolution: transition management in public policy. *Foresight*, 3(1), pp. 15–31.

Russell, S. and Williams, R., 2002. Social shaping of technology: frameworks, findings and implications for policy. In K.H. Sørensen and R. Williams (eds) *Shaping Technology, Guiding Policy: Concepts, Spaces & Tools*, pp. 37–131.

Seyfang, G. and Longhurst, N., 2013. Desperately seeking niches: grassroots innovations and niche development in the community currency field. *Global Environmental Change*, 23(5), pp. 881–891.

Shove, E., and Walker, G., 2007. CAUTION! Transitions ahead: politics, practice, and sustainable transition management. *Environment and Planning A*, 39(4), pp. 763–770.

Shue, V., and Thornton, P. (eds), 2017. *To Govern China: Evolving Practices of Power*. Cambridge: Cambridge University Press.

Sismondo, S., 2004. *An Introduction to Science and Technology Studies*. Oxford, Carlton: Blackwell.

Skjølsvold, T.M., 2014. Back to the futures: retrospecting the prospects of smart grid technology. *Futures*, 63, pp. 26–36.

Smith, A., Stirling, A., and Berkhout, F., 2005. The governance of sustainable socio-technical transitions. *Research Policy*, 34(10), pp. 1491–1510.

Sovacool, B.K. and Ramana, M.V., 2015. Back to the future: small modular reactors, nuclear fantasies, and symbolic convergence. *Science, Technology, & Human Values*, 40(1), pp. 96–125.

Star, S.L., 1991. Power, technologies and the phenomenology of conventions: on being allergic to onions. In J. Law (ed.) *A Sociology of Monsters: Essays on Power, Technology and Domination*. London and New York: Routledge, pp. 26–57.

Suttmeier, R.P. and Yao, X., 2004. *China's Post-WTO Technology Policy: Standards, Software, and the Changing Nature of Techno-Nationalism*. NBR Special Report no. 7.

Tyfield, D., Ely, A., and Geall, S., 2015. Low carbon innovation in china: from overlooked opportunities and challenges to transitions in power relations and practices. *Sustainable Development*, 23(4), pp. 206–216.

Unruh, G.C., 2000. Understanding carbon lock-in. *Energy Policy*, 28(12), pp. 817–830.

van Lente, H., 2000. Forceful futures: from promise to requirement. In N. Brown, B. Rappert and A. Webster (eds) *Contested Futures: A Sociology of Prospective Techno-Science*. London: Ashgate Publishing.

van Lente, H., 2012. Navigating foresight in a sea of expectations: lessons from the sociology of expectations. *Technology Analysis & Strategic Management*, 24(8), pp. 769–782.

Voß, J.P. and Bornemann, B., 2011. The politics of reflexive governance: challenges for designing adaptive management and transition management. *Ecology and Society*, 16(2).

Voß, J.P. and Kemp, R., 2005. Reflexive governance for sustainable development—incorporating feedback in social problem solving. Paper for European Society for Ecological Economics (ESEE) Conference, special session on Transition Management, 14–17 June, Lisbon. Available at: http://kemp.unu-merit.nl/pdf/Voss-Kemp%20Reflexive%20Governance%20for%20ESEE%202005.pdf (Accessed 3 November 2015).

Xue, L. and Liang, Z., 2010. Chapter 10. Relationships between IPR and technology catch-up: some evidence from China. In H. Odagiri and R.R. Nelson (eds) *Intellectual Property Rights, Development, and Catch-Up: An International Comparative Study*. New York: Oxford University Press.

3 Energy governance and innovation in China

Governance and politics in China can be difficult to understand for both insiders and outsiders. Few people know exactly what happens at the very top of the Chinese Communist Party (CCP), and those who know do not write books, especially not books in English. The country's policymaking has been described by Western academics as a 'maze of intricacies, complexities and contradictions' (Saich 2011, p. 4), or, more recently, as 'a complex of shifting forces, one most definitely with "no single hand" in control' (Shue and Thornton 2017, p. 13). The development China has experienced over the past 40 years challenges 'conventional wisdom as well as conventional models of political change' (Heilmann 2018, p. 1), because it cannot simply be placed somewhere between a plan and a market, or between democracy and dictatorship. Although complex, there are some things we can say about the way in which China is governed that are useful as a background for this book. Perhaps surprisingly, these things are not only descriptions of a well-planned system controlling processes from beginning to end, but also descriptions of what Ang (2016) has termed 'directed improvisation', where exerting *influence* is more important than exerting *control*. Thus, the story which will be foregrounded in this book is about recognising that, even in authoritarian regimes, there are uncertainties about the best ways to proceed, and that understanding what China has achieved since 1978 until today is about understanding governing practices as unfolding co-creative processes which appear as much in the moment at local levels as in central government plans and directives. These processes lay the foundations on which the development of China's wind and solar energy industries are built.

China's story of economic catch-up and transition towards a kind of market economy starts in the reform period initiated under Deng Xiaoping in 1978 (Saich 2011; Naughton 2007). The transition began gradually during the 1980s and was stepped up during the 1990s, especially after Deng's infamous 'Southern Tour' in 1992, during which he affirmed that the continued legitimacy for the CCP was dependent on economic growth (Saich 2011). Deng made this clear by stating that reform should not 'proceed slowly like a woman with bound feet' but should rather continue swiftly and 'blaze new trails boldly' (cited in Saich 2011, p. 86). Decentralisation and privatisation were important ingredients in the reform process (Green and Liu 2005; Naughton 2007). Indeed, the increased authority and independence allocated to local and provincial governments under Deng Xiaoping is claimed to have been essential to achieving the high economic performance and ensuring the legitimacy of the Communist Party (Zheng 2010). Moreover, the private sector has grown to become central to innovation activities. In 2005, the private sector accounted for more than two-thirds of China's gross domestic product and employment (Liu 2009; Saich 2011), and in 2012 the figure was estimated to be around 70 percent (Lardy 2019).

Another important ingredient of the rapid growth was the gradual opening of new special economic zones which functioned as testing ground for new policies, as well as important political symbols of change (Heilmann 2008b; Naughton 2007). The most known example is perhaps the Shenzhen special economic zone which opened in 1979 and grew from being a small market town to a large city of 12 million today. Since 1980 the number of these special zones has increased, and as of 2017, there were 279 different experimental zones with special economic legislation (Heilmann 2018, p. 85). It has increasingly become acknowledged that governance in China is not only a result of planning and control but is also a complex system in which uncertainty is embraced as a resource for change (Heilmann 2009; Heilmann and Perry 2011; Heilmann 2018). Indeed, China's history with heavy government involvement and control under socialist planning was not as large an obstacle to regime change as was expected by spectators (Naughton 2007; 2010). Instead of removing government institutions completely, as can be said to have happened in the Soviet transformation, China made use of government bodies which were seen to be 'reasonably well functioning organizations possess[ing] pools of valuable organizational and human capital' (Naughton 2010, p. 455). Along with this re-purposing of Chinese institutions went 'incentivization' of government, implying that government officials received bonuses or promotions depending on how well they fulfilled certain criteria (Naughton 2010). Most of these criteria were linked to economic growth.

Many China scholars have tried to explain how China's transformation has been possible, all whilst it has been based on out-dated, or even prehistoric, principles of communism. Some have argued that China soon will become a democracy (Gilley 2004), that the economy will collapse (Chang 2010), regress (Pritchett and Summers 2014), or that China is a 'fragile superpower' (Shirk 2007). It will not be the aim of this book to provide a grand theoretical contribution to this more overarching debate of Chinese governance. Rather, the aim of this book is to add a contribution from the specific experiences of the wind and solar industries to the understanding of Chinese politics as unfolding rather than only top-down controlled. This understanding is underpinned by the thesis set out by Heilmann and Perry (2011, p. 3) that the creative adaptation which has characterised Chinese governance recently has roots in China's revolutionary heritage, particularly in Maoist traditions of policy creation and implementation:

> China's Soviet-inspired formal institutions are combined with distinctive governance methods shaped by the Chinese Communists' own revolutionary and post-revolutionary past and, in the post-Mao era, complemented by selective borrowing from "advanced" foreign organizational and regulatory practices.

What is more, given the previous chapter's insistence on co-production it makes more sense to talk about Chinese governing *practices*, as opposed to institutions, since there is a danger which institutions remain *presumed* rather than actually demonstrated (Shue and Thornton 2017, p. 8). For this reason, it is useful to move 'beyond static authority relationships and incentive structures' and instead aim to 'illustrate a dynamic process of evolving and sometimes competing priorities, with periods of experimentation and consolidation' (Heilmann 2018, p. 195). Such an understanding resonates well with the overall undertaking of this book and underpins the conceptualisation of the *arts of the state*, which this book will further develop.

Flexibility, learning and policy experiments

This book argues that Chinese governing institutions are unique and should be studied on their own terms, instead of assuming what they could or should be. Scholarly discussions of governance and politics in China often revolve around tensions between centralisation and decentralisation, plan and market, local and national levels, rural and urban environments, or industrial and agricultural settings (Dittmer and Liu 2006; Fewsmith 2010; Lieberthal 2004; Saich 2011). Observing tensions in China's *energy* sector in the 1980s, Lieberthal and Oksenberg (1988) developed the highly influential concept of 'fragmented authoritarianism'. Their main conclusion was that the Chinese energy policy process is protracted, disjointed and incremental. Furthermore, the fragmentation of authority creates inter-ministerial competition and disjointed policymaking, because respective ministries have a similar level of authority, but disparate goals. This means that any policy initiative or major project 'need[s] to acquire the active cooperation of many bureaucratic units that are themselves nested in distinct chains of authority' (Lieberthal and Oksenberg 1988, p. 22). However, despite the fragmentation, the central leadership of the Chinese political system is able to exert influence and push through directives (Lieberthal and Lampton 1992).

Since Lieberthal and Oksenberg's thesis was developed, it has been increasingly acknowledged that flexibility, learning and adaptation have been central to China's massive transition process, and China scholars have put more emphasis on institutional capacities. Mertha (2009) argues that governance in China has become even more pluralised than when Lieberthal and Oksenberg did their analysis and describes the role of so called nontraditional and nonstate 'policy entrepreneurs' that have taken advantage of China's fragmented governance and successfully pursued their agendas. In many ways, what we observe in China is a co-production of politics which can be described as a type of policy learning. For instance, Gu and Lundvall (2006) highlight the importance of policy learning for China's innovation performance and emphasise the benefits of a simultaneously centralised and decentralised system. Heilmann and Perry (2011) have termed China's governing method 'guerrilla policy style' with reference to the governing methods adopted during Mao's reign, thus explaining how Chinese governing institutions have been able to manage sudden change and uncertainty throughout the decades. The Chinese central government, for instance, employs this type of campaign-style policy pushing which reaches down to the local level when there is uneven policy implementation, or when local governments deliberately lag behind in implementing new directives (Kennedy and Chen 2018).

Apart from the campaign-style pushes, the CCP also employs other important tools to retain influence, such as controlling availability of financial assets through state-owned banks, as well as through taxes, regulations, licenses, permits or subsidies (Jones and Zou 2017). A rather important mechanism of influence is the *bianzhi* and *nomenklatura* systems, wherein the CCP appoints CEOs and presidents for state-owned enterprises (SOEs), provincial governors, local and central officials in the Party (Jones and Zou 2017; Kirkegaard 2019, p. 69). The CCP can through this system place officials from the Party in strategic positions, reshuffle CEOs of strategic SOEs and make sure that those in central positions do not get too much influence in a particular company (Downs and Meidan 2011). This was the case in 2011, when the CCP moved around the top executives of the three-largest national oil companies: the China National Offshore Oil Corporation (CNOOC), China National Petroleum Corporation (CNPC) and China Petrochemical Corporation (Sinopec). Su Shulin, the general manager of Sinopec was appointed acting governor of Fujian Province, Fu Chengyu, general manager of CNOOC became chairman of Sinopec, and Wang Yilin, general manager of CNPC was made chairman of CNOOC (Downs and Meidan 2011). In essence, it was a reshuffle of leadership, and Downs and Meidan (2011, p. 4), hypothesise the reshuffle was related to four issues: 'patronage; the revolving door between government and business in China; improving corporate governance; and lastly, managing competition between the NOCs (National Oil Companies)'. In sum, state ownership does not equal control, but appointing leadership enables some degree of influence, or what Jones and Zou (2017, p. 743) call 'weak oversight and ongoing struggles'.

Heilmann (2008a, 2008b, 2009) coined the concept of 'experimentation under hierarchy' to describe the process by which China's institutional structure has innovated and adapted alongside large-scale economic change. Heilmann (2008b, p. 3) writes that this adaptability is due to a practice of policy experimentation in China which 'precedes the enactment of many national policies'. In short, he explains that policy experimentation, by delegating responsibility to local officials, 'reduced the frictions and delays characteristic of top-level consensus-building and interagency accommodation, and helped to avoid protracted policy deadlock' (Heilmann 2008b, p. 21). Thus, by starting with a smaller policy area, and scaling up only when successful, policy experimentation was a useful way to gain consensus amongst top-level politicians, since it entailed placing the policy burden on local governments and attributing the national success to the given policy (Heilmann 2008b). In this way, the success of a policy could be attributed to the central government, whilst responsibility for unsuccessful policy experiments can be confined to the local level (Heilmann 2008a; Heilmann 2008b; Naughton 2010). In this way, successful experiments can be scaled up without much internal disagreement, whilst failed experiments do not carry large consequences for the central policymakers who suggest them. China's decentralised structure therefore allows for experiments in selected areas, and successful experiments can then more easily be applied across the country (Andrews-Speed 2012). Part of the problem for the central government is therefore to *generate* consensus for political change, and experimentation can be considered one way in which the fragmentation of authority is lessened. Moreover, policy experiments are not only initiated from the top down in China. Policies can also be introduced from the bottom up, since fragmentation makes room for local governments to come up with own solutions. Overall, therefore experimental government policy, experience and practice have been important for fine-tuning China's policy machinery (Wang 2009). For this reason, we can also say that grassroots practices have been an important source of policy learning for the central government. These concepts of fragmentation, learning and adaptation are concepts which resonate well with the analysis of China's wind and solar industries.

Energy governance

Although it is nothing new to recognise that China's bureaucracy is fragmented, Mertha and Brødsgaard (2016) point out that China scholars need to constantly remind themselves and others who are less familiar with China of this fact. Looking closer at China's *energy* governance, it is established that the energy sector is still characterised by fragmented authoritarianism (Andrews-Speed 2012; Kong 2009). The very apex of the Communist Party, the standing committee—consisting of the seven-highest-ranked people in China—is the most important organ which can create consensus and implement decisions (Kong 2009; Naughton 2010; Andrews-Speed 2012; Lieberthal and Lampton 1992). Apart from the apex, however, there are internal power struggles between ministries, local and provincial governments, state-owned enterprises, and private entrepreneurs which were officially welcomed into the CCP in 2002 (Kennedy 2005; Dickson 2007; Saich 2011). These internal power struggles cause the decision-making process to drag out in an arduous process of bargaining, persuasion and arbitration, as consensus is needed to move forward (Saich 2011; Zheng 2010).

Governance of the energy sector is high-level politics in China. Li (2013) points out that all the members of China's Politburo Standing Committee, as well as several previous members, have important links to the energy sector. Many of them have either made a political career through the oil and gas industry or have been CEOs of some of China's largest oil companies. This, for instance, is clear from the example of the leadership shuffle in the national oil companies described above. Another case in point is former premier between 1988 and 1998 Li Peng and his family who have several leadership positions within oil and coal companies in China (Isoaho, Goritz and Schulz 2017). Li's children, Li Xiaolin and Li Xiaopeng, became the heads of China Power International (CPI) and Huaneng (two large energy SOEs), respectively, in the early 2000s; and in 2012, Li Xiaopeng became governor of the coal-rich Shanxi region (Isoaho, Goritz and Schulz 2017). Then, in 2016, Li Xiaopeng was promoted to Minister of Transport (reuters.com 2016). Energy decisions are particularly politicised in China because they involve many different actors with diverging interests and objectives. This has led to a state of affairs where there is

> A leadership vacuum in China over energy policy and many decisions are driven by projects promoted by localities or industries rather than being guided by a coherent national energy policy.
>
> (Kong 2009, p. 791)

Therefore, a change in energy policy amongst top-level leadership does not necessarily equate to smooth implementation throughout the system. Implementation can be constrained by vague and contradictory formulation of regulations, or because local government officials prioritise economic growth at the expense the environment (Meidan, Andrews-Speed and Xin 2009). Hence, even though power is centralised, this does not mean that it is concentrated in the centre.

The most important body regulating renewable energy policy centrally in China is the National Development and Reform Commission (NDRC). Within the NDRC, the National Energy Administration (NEA), established in 2008, is responsible for developing and implementing renewable energy policies. A problem the NEA faces, however, is that the administration is too heavily understaffed to be completely on top of emerging developments (Downs 2008b). Provincial governments also have their own Development and Reform Commissions which develop and implement local policies, sometimes in conflict with policies developed centrally (Mah and Hills 2014). This was, for instance, the case in the period between 2014 and 2016 when provincial level governments approved a large amount of new coal power plants following a regulatory devolution from central to provincial authorities (Shearer, Yu and Nace 2018). From March 2016 onwards, the central government started procedures to halt the construction of new plants, for instance by ordering construction of specific projects to slow down, but it appears that these measures were unsuccessful as of mid-2018 (Shearer, Yu and Nace 2018). Other important government actors are the 'big five' state-owned power generation companies, and the two grid companies State Grid and Southern Grid. The grid companies and the 'big five' have much influence in terms of what projects are developed and where, and often pursue their own agendas (Andrews-Speed 2012; Rosen and Houser 2007).

Until 2015, when the aforementioned 'Document No. 9' was introduced, the main concern for the Chinese government was to make sure enough electricity was produced. Between 2001 and 2006, electricity consumption grew on average by almost 14 percent annually, a staggering challenge to any power sector (Alva and Li 2018). At the end of 2017, China's power sector was more than 60 percent larger than the United States' power sector, but in the year 2000, it was less than one-third the size (Alva and Li 2018). In this period, therefore, there was a larger focus on ensuring enough investment in capacity growth rather than focusing on economic efficiency or environmental aspects (Alva and Li 2018). To make investments in coal power plants interesting, the 'fair-dispatch' and equal hour planning system was implemented so that each investor could gain cost recovery (Ho, Wang and Yu 2017). Simplified put, the way the system works is that a total number of annual generation hours is decided for each generator by the provincial government (Ho, Wang and Yu 2017). Grid companies then in turn have to ensure that each generator's year-end utilisation hours correspond to the quota they have been given (Ho, Wang and Yu 2017). Thus, since a newer and more efficient coal power plant is given the same number of operating hours as an older and less efficient one, the system gives inadequate consideration to efficiency or emissions related to each coal power plant (Wei *et al.* 2018). Today, therefore, China is facing serious overcapacity problems in coal-fuelled power generation (Wei *et al.* 2018). Wei *et al.* (2018, p. 337) calculated that the total overcapacity of coal generators in China equals to 41 percent of the total coal capacity. All these coal generators fight to sell as much power as possible, as the break-even running time of an average coal-fuelled power plant is 5,500 hours annually, which is higher than the average allocated running time of most coal-fuelled power plants (Wei *et al.* 2018). The problem is also that due to the overcapacity, each coal power plant is also used less—since 2015 average utilisation rates have dropped to 49 percent, meaning that the plants are idle for more time than they are in use (Shearer, Yu and Nace 2018). This is just one aspect of the several challenges which China's electricity sector reform process is currently facing.

Fragmentation prevails in China's energy sector, but Cunningham (2009) shows how the government uses liberalisation and consolidation as a means to control the growth in the coal and electric power industries, in what he terms a 'portfolio approach to energy governance'. Cunningham finds that central ownership of the electric power industry has fluctuated over time, demonstrating less regulation in times of electricity supply shortage, and more in times of sufficient electricity supply. This has led to periodisation of rapid expansions followed by contraction. Indeed, an alternating wave of consolidation and liberalisation has, over time, characterised China's electric power facilities, depending on the central government's concern at the moment. As we shall see in the next chapter, something similar can be said about fragmentation and centralisation in the wind and solar industries.

The role of state-owned enterprises (SOEs)

Another source of China's fragmentation of authority and influence are the state-owned enterprises (SOEs), which have gained increased autonomy through the period characterised by decentralisation, deregulation and reform since 1978 (Jones and Zou 2017). Especially in the energy sector, SOEs play an important role in policymaking and retain considerable influence in energy decision-making processes (Zheng 2010; Andrews-Speed 2012). One prime example is what is today known as State Grid (or the State Grid Corporation of China). In 1998, the Ministry for Electrical Power was abolished and replaced by the State Power Corporation, which, in 2002, was further unbundled into two grid companies, State Grid and Southern Grid, and five power generation companies referred to as 'the big five', Huaneng, Datang, Huadian, Guodian and State Power Investment Corporation (SPIC) (until 2015 known as China Power Investment Corporation) (Andrews-Speed 2012; Yi-chong 2012). In 2017, Guodian was merged with the Shenhua Group, a large state-owned coal producer, to become China Energy Investment Corporation, the world's largest energy utility (reuters.com 2017). Although State Grid has been fostered by and has had its 'umbilical cord' attached to the government, it has 'developed a life of its own' (Yi-chong 2012, p. 129), and has ranked seventh in the Global Fortune 500 every year since 2011 (Yi-chong 2017). Today, State Grid in the world's largest utility, employs about 1.8 million people and provides electricity to about 1.1 billion people covering 88 percent of China's territory (the rest is covered by China Southern Grid) (Yi-chong 2017).

State Grid is one of the key strategic centrally administered SOEs and is directly managed by the State-owned Assets Supervision and Administration Commission of the State Council (SASAC). As of 2019, SASAC administered 96 SOEs (commonly referred to as 'central SOEs'), many of which concentrated on monopoly industries such as oil and gas, electricity or telecommunication (Zheng 2010; SASAC 2019). Because there were an increasing number of loss-making SOEs in China throughout the 1990s, the central government initiated a reorganisation which involved privatising smaller SOEs and encouraging others to merge into larger SOEs. This policy was known as 'grabbing the big and letting go of the small' (Yi-chong 2017, p. 7). SASAC was established in 2003 to assist with this process and started out with a total of 196 government owned companies under its jurisdiction (Yi-chong 2017). At the time, 40 percent of these central SOEs were insolvent, and a major task was to reform them; restructuring them and transforming them into modern corporations (Yi-chong 2017). Therefore, several of these SOEs were merged and the total number of central SOEs has decreased. SASAC is not a direct owner, however, but should be considered more as a shareholder with the main aim of increasing the profitability of the central SOEs (Jones and Zou 2017). In fact, 63 percent of SASAC's portfolio is listed on the stockmarket (economist.com 2018). Nevertheless, SASAC was not the only government agency responsible for dealing with the large SOEs, and SASAC had to 'negotiate, bargain, compete, and cooperate' with these other agencies that did not always have aligned visions and interests (Yi-chong 2017, p. 96).

State Grid has been described as a 'Frankenstein's monster' with growing political influence and an increasing amount of resources at its disposal, 'eventually terrorising its own creator' (Yi-chong 2012, p. 147). In other words, the company has become so large that an increasing amount of resources, both sticks and carrots, are needed for the government to manage it well. State Grid is a key actor in the story of the energy transition to more renewable energy in China—it can even be said that State Grid action is a central premise for the success or failure of this transition. It is in China's interest to provide stable, reliable and affordable electricity to its citizens. That this electricity on top of this should be sustainable is a major challenge currently facing the State Grid.

Renewable electricity is unstable, intermittent and (until recently) more expensive than electricity from coal and hydro power plants. This is something which poses considerable challenges for State Grid, because wind and solar energy require more flexible grid management systems than coal power (Yi-chong 2017). The integration of wind and solar electricity requires a change of routines of the way in which dispatch centres are run and finding incentives to do this might not be too easy for local managers. A story from an interviewee in one of the grid companies (Interview #50, April 2014) points this out. He visited a grid operating utility in Inner Mongolia, a province with a very high penetration of wind energy, and saw a note on the wall saying, 'Always curtail wind power'. This attitude to curtailing wind was connected to a general habit of not planning the dispatch because with coal power it is much easier. He added that accurate forecasting and implementation of the forecast is crucial to optimise electricity dispatch, and that 'curtailment is related to many issues that are inherent in power system dynamics: load, grid/transmission, and the resource at hand'. We will get back to this later; however, this is just one out of several challenges facing State Grid. As pointed out by Yi-chong (2017, p. 41), State Grid must 'deal with a difficult mix of challenges that include a continuing (if slower) rise in electricity demand; limited natural energy resources; a mismatch of locations between resources and electricity demand; and worsening air, water, and soil pollution'.

As the case of State Grid shows, 'state-owned' does not equate to 'state-controlled'. Instead, many of these powerful central SOEs are relatively autonomous actors, and sometimes advance corporate ahead of national interests (Downs 2008b). As Downs (2008a, p. 137) puts it: 'China's leadership faces some of its greatest new challenges in the energy sector partly because the projects pursued by the energy SOEs tend to shape the country's energy policies rather than vice versa'.

Innovation in China

The topic of institutional change in China is big, and it is often linked to discussions around innovation, economic growth and central-local relationships and friction (Teets 2015). At the World Economic Forum in Davos in January 2015, China's premier, Li Keqiang, indicated that the 'new normal' for China's economic growth is 'shifting from high speed to medium-to-high speed, and development needs to move from low-to-medium level to medium-to-high level' (weforum.org 2015). Over the past three decades, the two most important factors in China's rapid economic growth have been external demand combined with increased investment from both domestic and foreign sources (Wu 2006). Foreign direct investment has, since the beginning of the 1980s, been an important method of attracting technology to China and increasing exports (Long 2005; Naughton 2007). It is believed that foreign investment may contribute technologically in three ways: filling technological gaps, introducing advanced technology and improving existing technology (Long 2005). However, since foreign companies own the patents, copyrights and trademarks, they also receive the most profit from the technology (Kennedy 2010). In 2009, 56 percent of China's exports were produced in foreign invested companies, and in high-tech exports the share was as high as 85 percent (Kennedy 2010). One challenge for Chinese companies has been to build local competencies whilst competing with larger and more advanced foreign companies. The Chinese government surely expected technology to 'spill over' so as to establish domestic companies, as occurred in South Korea (Kim 1997). However, the spillover in the high-technology sector has been much lower than expected (Moran 2011; Xiao, Tylecote and Liu 2013), hence Li Keqiang's comment on moving from low to high levels of development. China no longer wants to be the world's manufacturing floor but wants to contribute to global innovation; in short, China wants to shift from 'made in China' to 'designed in China' (Cao, Suttmeier and Simon 2006). How can this be achieved?

There is a large and ongoing debate on innovation and the role of government in China. To simplify somewhat, I group the debate into those who, on the one hand, do not believe that innovation is occurring in China, and that this is due to a cultural or systemic component or to government involvement. On the other hand, there are those who believe that innovation is happening, but largely in private firms (Nee and Opper 2012), or because the government can function as a 'state entrepreneur' that facilitates innovation (Shi, Markóczy and Stan 2014; Sun 2015). Several of these stances can also be combined (Yang *et al.* 2012), for instance by saying that a certain type of innovation is happening in private firms, but that government control and interference is stifling another type of innovation. This was an argument made by Breznitz and Murphree (2011), who looked at China's telecommunication companies and found that government interference hindered companies from novel-product innovation. This happens, they claim, since

> top managers are very much aware that if all goes well and they do not make any grave mistakes, they will likely be promoted and transferred to one of their rivals within a couple of years, they have little incentive for extensive investment in novel-technology, which, although it might give their current company an edge or even propel it to market dominance, carries high personal risk.
>
> (Breznitz and Murphree 2011, p. 58)

Breznitz and Murphree (2011) argue that growth and 'manufacturing-innovation' in China have come about largely *because* of central-local fragmentation and not as a result of central government efforts. 'Novel-product' innovations have largely failed to materialise, they argue, because the Chinese system gives preference to short-term and tangible profit. Agreeing that state support has not been central to innovation in China, Nahm and Steinfeld (2014, p. 288) contend that China has become a 'scale up nation', where firms have developed 'unique capabilities surrounding technology commercialization and manufacturing-related innovation'. A similar argument is sustained by Wan *et al.* (2015), who claims that China is a particularly fertile environment for *disruptive* innovations, such as 'cost innovation, application innovation, and business model innovation'. Much research thus claims that what is happening in China is not novel-product innovation, but rather disruptive and manufacturing innovation, and that the government has not played an important role in ensuring that type of innovation.

This view is somewhat contrasted by scholars who believe that innovation is happening in China (e.g., Rein 2014), and that an important source of innovation is state coordination and strategic governance (Heilmann 2010; Steinfeld 2010). Steinfeld (2010, p. 164), for instance, claims that 'whether they are multinational or indigenous firms doing the work, it is in the China market that numerous new-to-the-world energy technologies are being developed and deployed for the first time'. According to Steinfeld, this is particularly true for nuclear and thermal electric power generation technologies. Moreover, analysts of the 'entrepreneurial state' hold that the courage, foresightedness and stability of the government have been crucial for paving the way for new technological developments, such as renewable energy technologies (Mazzucato 2013). Sun (2015) adapts this view to the Chinese government and finds that alertness to opportunities, resource exploration and consolidation and strategic learning are key components of China's entrepreneurial state. Using high-speed trains as a case in point, Sun claims that these three components have been crucial to building new innovative technologies in China. Nevertheless, developing a 'pessimist' stance on innovation in China, Tyfield (2017, p. 94), remarks that 'China is almost the very antithesis of the "entrepreneurial state"', because of the unfortunate combination of both 'too much' and 'too little' state. The gist of the argument is that there is simply too much haphazardness, uncertainty and complexity for management which disincentivises long-term innovation investment.

On the 'doubting' side, there are also strong opinions regarding the influence of culture on innovation performance in China. President Xi Jinping himself, for instance, is not happy with the innovation efforts in China. As he has put it, in rather poetic language: 'We cannot always decorate our tomorrows with others' yesterdays. We cannot afford to lag behind in this important race. We must catch up and then try to surpass others' (quoted in Economy [2018, p. 122]). After the death of Steve Jobs in 2011, a discussion surged in China regarding 'Why can China not produce its own Steve Jobs?' (e.g., chinadaily.com.cn 2011; Mourdoukoutas 2011). Experts from China and abroad provided explanations ranging from cultural to systemic issues. For instance, one observer wrote that 'China's traditional Confucianism cannot integrate with the Western Schumpeter corporate spirit, so throughout Chinese history, we rarely see respect for innovators and businessmen' (chinadaily.com.cn 2011). On a more institutional level, Gu *et al.* (2009, p. 385), in an overview article on innovation in China, state that

> Chinese people believe in the positive role of science and technology while there is a weakness in social capital—values, norms and trust that are shared by the social community. Weak social capital is reflected in the low level of willingness by either academia or business enterprises to interact, cooperate and engage in knowledge sharing with individuals and organizations outside their family circles or immediate 'neighbors'.

Social capital, as Gu *et al.* (2009) see it, is important in order for firms to learn and necessary for innovation and new ideas to occur. Furthermore, Serger (2009, p. 64) claims that the reason for this low willingness to share knowledge with outsiders in China is that there is a low level of low level of trust between firms and other market actors. Corruption and intellectual property rights (IPR) infringements are taken as proxies of the level of trust, with high levels representing low trust (OECD 2008). Another scholar, Baark (2007, p. 354), asserts in a study of knowledge and innovation institutions in China that

> the predicament of innovators [in China] today may still be difficult, swimming as they do in a sea of acidic attitudes towards the new and untried, with precious little effective support from peers, officials, or the population at large.

This quote alludes to China's education system, which, according to Baark (2007, p. 354), still appears to be too heavily influenced by legacies of reproducing and refining the knowledge of the great masters of the past, encouraging 'imitation more than invention'.

In a thorough review of innovation policy in China, Teets (2015) attempts to understand why government officials are willing to engage in relatively 'risky' innovation attempts, when simpler—and more directly growth-oriented—choices could have been made. Her conclusion is that we still 'lack an explanation for why local officials engage in this risky behaviour' (Teets 2015, p. 104). Nevertheless, she argues, if we could understand better the motivation of local officials engaging in innovation, we could also gain better understandings of the way in which 'subnational experimentation represents a path to successful political and economic change without fundamental (and politically infeasible) reform of governing institutions' (p. 105). Tyfield (2017) points out that a crucial element to understand innovation in China is precisely the possibilities which arise due to elements identified in both the 'believing' and the 'doubting' sides of the innovation discussion. Nevertheless, it is premised on a constant and dynamic search for incremental advantages:

> This is thus an essentially dynamic innovation model that instantiates and feeds, in positive feedback loops, both the existing 'strengths' and 'weaknesses', as conventionally defined, of the Chinese innovation system and their increasingly intense interaction that is the dynamism of the system's evolution itself.
>
> (Tyfield 2017, p. 109)

In general, from the above discussion we can conclude that the answer to whether or not innovation is happening in China depends strongly on the definition of innovation employed. There are, therefore, scholars who believe that a particularly Chinese view of innovation must be interpreted through its own logic, and not through the same lens used to view Western innovation (Vinig and Bossink 2015). For this reason, Vinig and Bossink (2015, p. 621) propose that 'to better understand Chinese indigenous innovation, a new paradigm of Chinese innovation should be developed'. In any case, it is clear that innovation is a process which cannot be controlled, and both intended and unintended consequences of innovation efforts appear to be central to achieving successful innovation.

The CCP firmly attempts to make sure that innovation happens in China, and that the country will become a leading innovation nation. This is manifested in China's 'indigenous innovation' programme, expressed in the 15-year 'Medium to Long-Term Plan for the Development of Science and Technology' (MLP), issued in 2006, as well as the more recent 'Made in China 2025 strategy'. The 2006 MLP plan states that China shall transform into an 'innovation-orientated nation' by 2020, and details how China will deal with issues such as technological dependency and domestic innovation, environmental concerns and resource utilisation, as well as human resources and education (Cao, Suttmeier and Simon 2006; Cao, Suttmeier and Simon 2009). The plan has a stated goal of increasing the national expenditure on research and development from a base of 1.34 percent of gross domestic product in 2005 to 2.5 percent in 2020, and to reduce the dependence on imported technology to no more than 30 percent (Cao, Suttmeier and Simon

2009). Another aspect of the Chinese 'indigenous innovation' programme is not only related to technological independency, but also to developing international standards to enhance national power and international influence (Kennedy, Suttmeier and Su 2008; Ernst 2011). Thus far, however, it appears that China's success in developing domestic standards has been limited (Kennedy, Suttmeier and Su 2008; Fomin, Su and Gao 2011; Gao and Liu 2012). Standards are important for economic performance, since the 'owners' of a standard can shape the direction of technological developments, thus giving long-term advantages to early movers (Gao and Liu 2012; Fomin, Su and Gao 2011). As Suttmeier and Yao (2004, p. 17) point out, 'the technological winner is now the one who manages to control de facto market standards while at the same time protecting intellectual property rights'.

Perhaps because of the lacking success of previous plans, the ten-year strategy named 'Made in China 2025' was launched in 2015 with the aims of turning China into a 'manufacturing superpower' and 'internet superpower' within the next decade (Wübbeke *et al.* 2016). The plan is not focussing so much on 'indigenous innovation' or the seven 'strategic emerging industries' formulated in the 2006 MLP but sketches a rather broad plan encompassing manufacturing not only in advanced industries, but also in traditional and service industries (Kennedy 2015). The plan is inspired by the German Industry 4.0 approach for intelligent manufacturing industries with the aim to increase the domestic content of important components and materials, ultimately occupying 'the highest parts of global production chains' (Kennedy 2015). The plan specifies ten target industries which are thought to be at the very core of China's strategic interests over the next ten years. Amongst these are new advanced information technology, including artificial intelligence and quantum computing, but also modern rail transport equipment, self-driving and new-energy vehicles and power equipment (supchina.com 2018). Power equipment includes new-energy and renewable-energy equipment, such as that related to solar PV and wind energy. The plan went rather unnoticed during the first couple of years, but it has received a lot of attention lately, as it is one of the core reasons for the initiation of the 'trade war' with the United States in mid-2018 (supchina.com 2018). This, for instance, was evident as several of the Chinese goods which were targeted for a 25 percent tariff by the United States in July 2018 were related to the Made in China 2025 plan, such as aerospace, automobiles, information technology and robotics (Swanson 2018).

In my view, the reaction which the plan has spurred from the US side implies that it is getting more and more likely that China is actually getting closer to succeeding in developing domestic innovation capabilities, including within high-tech and cutting-edge industries. It is nonetheless not straightforward to understand exactly how China will achieve this. During the Great Leap Forward at the end of the 1950s, the Communist Party slogan was to 'overtake Britain and catch up with the US' (Hu 2013; Christensen 2013). Catching up with 'frontier' countries has been an important strategic achievement for China's development since the Communist Party came to power in

1949. It should therefore not come as a surprise to anyone that knowledge is political in China (Gorm Hansen 2017). However, just as it was a challenge for Japan and Korea before (Amsden 1989), it appears to have been a challenge for China to develop independent, 'indigenous' technological capabilities whilst, at the same time, wanting to learn from foreign countries. In order to get a better understanding of what it means to actually achieve advanced manufacturing and innovation domestically, it is easier to look closer at the strategies undertaken in particular industries. The next chapter therefore looks closer at the particular strategies chosen in the wind and solar industries. The foundations built to gain market shares in these industries started more than a decade before the 'Made in China 2025' plan was launched.

References

Alva, C.A.H. and Li, X. 2018. Power sector reform in China. An international perspective. Report published by the OECD/IEA, International Energy Agency.

Amsden, A.H., 1989. *Asias's Next Giant. South Korea and Late Industrialization*. New York, Oxford: Oxford University Press.

Andrews-Speed, P., 2012. *The Governance of Energy in China: Transition to a Low-Carbon Economy*. Houndmills, Basingstoke, Hampshire/New York: Palgrave Macmillan.

Ang, Y.Y., 2016. *How China Escaped the Poverty Trap*. Ithaca, NY: Cornell University Press.

Baark, E., 2007. Knowledge and innovation in China: historical legacies and emerging institutions. *Asia Pacific Business Review*, 13(3), pp. 337–356.

Breznitz, D. and Murphree, M., 2011. *Run of the Red Queen: Government, Innovation, Globalization, and Economic Growth in China*. New Haven, London: Yale University Press.

Cao, C., Suttmeier, R.P. and Simon, D.F., 2006. China's 15-year science and technology plan. *Physics Today*, (December), pp. 38–43.

Cao, C., Suttmeier, R.P. and Simon, D.F., 2009. Chapter 10. Success in state-directed innovation? Perspectives on China's plan for the development of science and technology. In G. Parayil and A.P. D'Costa (eds) *The New Asian Innovation Dynamics: China and India in Perspective*. Houndmills: Palgrave Macmillan.

Chang, G.G., 2010. *The Coming Collapse of China*. New York: Random House.

chinadaily.com.cn, 2011. Where is China's Steve Jobs? *China Daily*. Available at: www.chinadaily.com.cn/opinion/2011-10/14/content_13899857.htm.

Christensen, N.H., 2013. *Shaping Markets: A Neoinstitutional Analysis of the Emerging Organizational Field of Renewable Energy in China*. PhD thesis at the Doctoral School of Organisation and Management Studies at Copenhagen Business School. Available at: http://openarchive.cbs.dk/bitstream/handle/10398/8627/Nis_Høyrup_Christensen.pdf.

Cunningham, E.A., 2009. *A Portfolio Approach to Energy Governance: State Management of China's Coal and Electric Power Supply Industries*. Doctoral dissertation, Cambridge, MA: Department of Political Science, Massachusetts Institute of Technology.

Dickson, B.J., 2007. Integrating wealth and power in China: the Communist Party's embrace of the private sector. *The China Quarterly*, 192, pp. 827–854.

Dittmer, L. and Liu, G. (eds), 2006. *China's Deep Reform: Domestic Politics in Transition*, Lanham, Boulder, New York, Toronto, Oxford: Rowman & Littlefield.

Downs, E.S., 2008a. business interest groups in chinese politics: the case of the oil companies. In C. Li (ed.) *China's Changing Political Landscape: Prospects for Democracy*. Washington, D.C.: Brookings Institution Press, pp. 121–142.

Downs, E.S., 2008b. China's "new" energy administration: China's national energy administration will struggle to manage the energy sector effectively. *China Business Review*. Available at: www.brookings.edu/~/media/Files/rc/articles/2008/11_china_energy_downs/11_china_energy_downs.pdf.

Downs, E.S. and Meidan, M., 2011. Business and politics in China. *China Security*, 19, pp. 3–21.

economist.com, 2018. Are China's state giants reformable? Available at: www.economist.com/business/2018/03/01/are-chinas-state-giants-reformable (Accessed 13 March 2019).

Economy, E., 2018. *The Third Revolution. Xi Jinping and the New Chinese State*. Oxford University Press. Kindle Edition.

Ernst, D., 2011. *Indigenous Innovation and Globalization: The Challenge for China's Standardization Strategy*. A joint publication of the UC Institute on Global Conflict and Cooperation and the East-West Center, Honolulu, Hawaii.

Fewsmith, J. (ed.), 2010. *China Today, China Tomorrow: Domestic Politics, Economy, and Society*. Lanham, MD: Rowman & Littlefield.

Fomin, V.V., Su, J. and Gao, P., 2011. Indigenous standard development in the presence of dominant international standards: the case of the AVS standard in China. *Technology Analysis & Strategic Management*, 23(7), pp. 745–758.

Gao, X. and Liu, J., 2012. Catching up through the development of technology standard: the case of TD-SCDMA in China. *Telecommunications Policy*, 36(7), pp. 531–545.

Gilley, B., 2004. *China's Democratic Future: How It Will Happen and Where It Will lead*. New York: Columbia University Press.

Gorm Hansen, L.L., 2017. Triggering earthquakes in science, politics and Chinese hydropower—a controversy study. PhD Thesis at the Copenhagen Business School. Available at: www.cbs.dk/en/research/departments-and-centres/department-of-international-economics-and-management/events/phd-defence-louise-lyngfeldt-gorm-hansen-0 (Accessed 11 March 2019).

Green, S. and Liu, G.S. (eds), 2005. *Exit the Dragon? Privatization and State Control in China*, London: Blackwell Publishing Inc.

Gu, S., and Lundvall, B.-Å. 2006. China's innovation system and the move toward harmonious growth and endogenous innovation. DRUID Working Paper No. 06-7. Danish Research Unit for Industrial Dynamics (DRUID). Available at: http://www3.druid.dk/wp/20060007.pdf (Accessed 11 September 2011).

Heilmann, S., 2008a. From local experiments to national policy: the origins of China's Distinctive policy process. *The China Journal*, 59, pp. 1–30.

Heilmann, S., 2008b. Policy experimentation in China's economic rise. *Studies in Comparative International Development (SCID)*, 43(1), pp. 1–26.

Heilmann, S., 2009. Maximum tinkering under uncertainty: unorthodox lessons from China. *Modern China*, 35(4), pp. 450–462.

Heilmann, S., 2010. Chapter 6. Economic governance: authoritarian upgrading and innovative potential. In J. Fewsmith (ed.) *China Today, China Tomorrow: Domestic Politics, Economy, and Society*. Lanham, MD: Rowman & Littlefield, pp. 109–129.

Heilmann, S. and Perry, E.J., 2011. Embracing uncertainty: guerrilla policy style and adaptive governance in China. In S. Heilmann and E.J. Perry (eds) *Mao's Invisible Hand: The Political Foundations of Adaptive Governance in China*. Harvard University Press, pp. 1–29.

Heilmann, S., 2018. *Red Swan: How Unorthodox Policy Making Facilitated China's Rise.* Hong Kong: The Chinese University Press.

Ho, M.S., Wang, Z. and Yu, Z., 2017. China's power generation dispatch. Report prepared for Resources for the Future (RFF).

Hu, A., 2013. *Political and Economic History of China, Volume 2: From Great Leap Forward to Reconstructing the Economy (1957–1965)*. Singapore: Enrich Professional Publishing.

Isoaho, K., Goritz, A. and Schulz, N., 2017. Chapter 12: governing clean energy transitions in China and India. In D. Arent, C. Arndt, M. Miller, F. Tarp and O. Zinaman (eds) *The Political Economy of Clean Energy Transitions*. Oxford, UK: Oxford University Press, pp. 231–250.

Jones, L. and Zou, Y., 2017. Rethinking the role of state-owned enterprises in China's rise. *New Political Economy*, 22(6), pp. 743–760.

Kennedy, S., 2005. *The Business of Lobbying in China*. Cambridge, MA: Harvard University Press.

Kennedy, S., 2010. Indigenous innovation: not as scary as it sounds. *China Economic Quarterly*, 14(3), pp. 15–20.

Kennedy, S. 2015. Made in China 2025. Available at: www.csis.org/analysis/made-china-2025 (Accessed 11 March 2019).

Kennedy, S., Suttmeier, R.P., Su, J., 2008. Standards, stakeholders, and innovation—China's evolving role in the global knowledge economy. NBR Special report no. 15, The National Bureau of Asian Research, Seattle, Washington.

Kennedy, J.J. and Chen, D., 2018. State capacity and cadre mobilization in China: the elasticity of policy implementation. *Journal of Contemporary China*, 27(111), pp. 393–405.

Kim, L., 1997. *Imitation to Innovation: The Dynamics of Korea's Technological Learning*, Boston: Harvard Business School Press.

Kirkegaard, J.K., 2019. *Wind Power in China. Ambiguous Winds of Change in China's Energy Market*. Abingdon and New York: Routledge.

Kong, B., 2009. China's energy decision-making: becoming more like the United States? *Journal of Contemporary China*, 18(62), pp. 789–812.

Lardy, N.R., 2019. *The State Strikes Back: The End of Economic Reform in China?*. Peterson Institute for International Economics.

Li, X.F., 2013. Chinese political transition makes change possible in energy approach. *Oil & Gas Journal*, 111(2), pp. 46–46.

Lieberthal, K., 2004. *Governing China: From Revolution Through Reform*. New York, London: W.W. Norton and Company.

Lieberthal, K. and Lampton, D.M. (eds), 1992. *Bureaucracy, Politics, and Decision Making in Post-Mao China*. Berkeley, Los Angeles, Oxford: University of California Press.

Lieberthal, K. and Oksenberg, D.M., 1988. *Policy Making in China—Leaders, Structures, and Processes*, Princeton, NJ: Princeton University Press.

Liu, X. 2009. Chapter 5: national innovation systems in developing countries: the Chinese national innovation system in transition. In B.-Å. Lundvall, K.J. Joseph,

C. Chaminade and J. Vang (eds), *Handbook of Innovation Systems and Developing Countries: Building Domestic Capabilities in a Global Setting*. Cheltenham: Edward Elgar Publishing.

Long, G., 2005. Chapter 12: China's policies on FDI: review and evaluation. In T.H. Moran, E.M. Graham and M. Blomström (eds) *Does Foreign Direct Investment Promote Development?* Washington, D.C.: Center for Global Development, Institute for International Economics.

Mazzucato, M., 2013. *The Entrepreneurial State: Debunking Public vs. Private Myths in Risk and Innovation*. London, New York, Delhi: Kindle Edition.

Meidan, M., Andrews-Speed, P. and Xin, M., 2009. Shaping China's energy policy: actors and processes. *Journal of Contemporary China*, 18(61), pp. 591–616.

Mertha, A., 2009. "Fragmented authoritarianism 2.0": political pluralization in the Chinese policy process. *The China Quarterly*, 200(December), pp. 995–1012.

Mertha, A. and Brødsgaard, K.E., 2016. Introduction. In Brødsgaard, K.E. (ed.). *Chinese Politics as Fragmented Authoritarianism: Earthquakes, Energy and Environment*. London and New York: Taylor & Francis.

Moran, T.H., 2011. *Foreign Manufacturing Multinationals and the Transformation of the Chinese Economy: New Measurements, New Perspectives*, Washington: Peterson Institute for International Economics. Available at: www.piie.com/publications/wp/wp11-11.pdf.

Mourdoukoutas, P., 2011. Why China doesn't have its Own Steve Jobs. *forbes.com*. Available at: www.forbes.com/sites/panosmourdoukoutas/2011/10/09/why-china-doesnt-have-its-own-steve-jobs/.

Nahm, J. and Steinfeld, E.S., 2014. Scale-up nation: China's specialization in innovative manufacturing. *World Development*, 54 (February), pp. 288–300.

Naughton, B., 2007. *The Chinese Economy. Transitions and Growth*. Cambridge, Mass.: MIT Press.

Naughton, B., 2010. China's distinctive system: can it be a model for others? *Journal of Contemporary China*, 19(65), pp. 437–460.

Nee, V. and Opper, S., 2012. *Capitalism from Below: Markets and Institutional Change in China*. Cambridge, MA: Harvard University Press.

Ngar-yin Mah, D. and Hills, P.R., 2014. Policy learning and central–local relations: a case study of the pricing policies for wind energy in China (from 1994 to 2009). *Environmental Policy and Governance*, 24(3), pp. 216–232.

OECD, 2008. *OECD Reviews of Innovation Policy: China 2008*. OECD Publishing. Available at: /content/book/9789264039827-en. Online. ISSN: 19934211 (online) https://doi.org/10.1787/19934211.

Pritchett, L. and Summers, L.H., 2014. *Asiaphoria Meets Regression to the Mean*. NBER Working Paper No. 20573. National Bureau of Economic Research (NBER).

Rein, S., 2014. *The End of Copycat China: The Rise of Creativity, Innovation, and Individualism in Asia*. Hoboken, NJ: Wiley. Kindle Edition.

reuters.com, 2016. Son of former Chinese premier promoted to transport minister. Available at: www.reuters.com/article/us-china-politics-idUSKCN11903K (Accessed 12 March 2019).

reuters.com, 2017. Factbox: Shenhua and Guodian—China's latest state marriage. Available at: www.reuters.com/article/us-china-power-shenhua-guodian-factbox-idUSKCN1B918I (Accessed 11 May 2019).

Rosen, D. and Houser, T., 2007. *China Energy: A Guide for the Perplexed.* Center for Strategic and International Studies and the Peterson Institute for International Economics. Available at: www.iie.com/publications/papers/rosen0507.pdf (Accessed 16 May 2011).

Saich, T., 2011. *Governance and Politics of China* (3rd ed.). Houndmills, Basingstoke, Hampshire/New York: Palgrave Macmillan.

SASAC, 2019. Directory. Available at: http://en.sasac.gov.cn/directorynames.html (Accessed 8 February 2019).

Serger, S.S., 2009. Chapter 3: foreign corporate R&D in China: trends and policy issues. In G. Parayil and A.P. D'Costa (eds) *The New Asian Innovation Dynamics: China and India in Perspective*. Houndmills: Palgrave Macmillan.

Shearer, C., Yu, A. and Nace, T., 2018. Tsunami warning. Can China's central authorities stop a massive surge in new coal plants caused by provincial overpermitting? Report by CoalSwarm. Available at: https://endcoal.org/2018/09/tsunami-warning/.

Shi, W., Markóczy, L. and Stan, C.V., 2014. The continuing importance of political ties in China. *Academy of Management Perspectives*, 28(1), pp. 57–75.

Shirk, S.L., 2007. *China: Fragile Superpower.* Oxford and New York: Oxford University Press.

Shue, V. and Thornton, P. (eds), 2017. *To Govern China: Evolving Practices of Power.* Cambridge: Cambridge University Press.

Steinfeld, E.S., 2010. *Playing our Game: Why China's Rise Doesn't Threaten The West.* Oxford: Oxford University Press.

Sun, Z., 2015. Technology innovation and entrepreneurial state: the development of China's high-speed rail industry. *Technology Analysis & Strategic Management*, 27(6), pp. 646–659.

supchina.com, 2018. Made in China 2025: the domestic tech plan that sparked an international backlash. Available at: https://supchina.com/2018/06/28/made-in-china-2025/ (Accessed 11 March 2019).

Suttmeier, R.P. and Yao, X., 2004. *China's Post-WTO Technology Policy: Standards, Software, and the Changing Nature of Techno-Nationalism.* NBR Special report #7, The National Bureau of Asian Research, Seattle, Washington.

Swanson, A., 2018. U.S. and China expand trade war as Beijing matches Trump's tariffs. Available at: www.nytimes.com/2018/06/15/us/politics/us-china-tariffs-trade.html (Accessed 11 March 2019).

Teets, J.C., 2015. The politics of innovation in China: local officials as policy entrepreneurs. *Issues and Studies*, 51(2), pp. 79–109.

Tyfield, D., 2017. *Liberalism 2.0 and the Rise of China: Global Crisis, Innovation and Urban Mobility.* London and New York: Routledge. Kindle Version.

Vinig, T. and Bossink, B., 2015. China's indigenous innovation approach: the emergence of Chinese innovation theory? *Technology Analysis & Strategic Management*, 27(6), pp. 621–627.

Wan, F., Williamson, P.J. and Yin, E., 2015. Antecedents and implications of disruptive innovation: Evidence from China. *Technovation*, 39–40, pp. 94–104.

Wang, S. 2009. Adapting by learning: the evolution of China's rural health care financing. *Modern China*, 35(4), 4, pp. 370–404.

weforum.org, 2015. Chinese Premier Li Keqiang's speech at Davos 2015. *World Economic Forum.* Available at: https://agenda.weforum.org/2015/01/chinese-premier-li-keqiangs-speech-at-davos-2015/ (Accessed 19 March 2015).

Wei, Y.M., Chen, H., Chyong, C.K., Kang, J.N., Liao, H. and Tang, B.J., 2018. Economic dispatch savings in the coal-fired power sector: an empirical study of China. *Energy Economics*, 74(August), pp. 330–342.

Wu, Y., 2006. *Economic Growth, Transition, and Globalization in China*, Advances in Chinese economic studies series. E. Elgar, Cheltenham.

Wübbeke J., Meissner, M., Zenglein, M.J., Ives, J. and Conrad, B., 2016. Made in China 2025. The making of a high-tech superpower and consequences for industrial countries. Report prepared for the Mercator Institute for China Studies (merics). Available at: www.merics.org/sites/default/files/2017-09/MPOC_No.2_MadeinChina2025.pdf (Accessed 11 March 2019).

Xiao, Y., Tylecote, A. and Liu, J., 2013. Why not greater catch-up by Chinese firms? The impact of IPR, corporate governance and technology intensity on late-comer strategies. *Research Policy*, 42(3), pp. 749–764.

Yang, J., Liu, H., Gao, S. and Li, Y., 2012. Technological innovation of firms in China: Past, present, and future. *Asia Pacific Journal of Management*, 29(3), pp. 819–840.

Zheng, Y., 2010. Chapter 11: central-local relations. The power to dominate. In J. Fewsmith (ed.) *China Today, China Tomorrow: Domestic Politics, Economy, and Society*. Lanham, MD: Rowman & Littlefield, pp. 193–223.

Yi-chong, X., 2012. The state grid corporation of China. In X. Yi-chong (ed.). *The Political Economy of State-owned Enterprises in China and India*. New York: Palgrave Macmillan, pp. 128–151.

Yi-chong, X., 2017. *Sinews of Power. The Politics of the State Grid Corporation of China*. New York: Oxford University Press.

4 The emergence of China's wind and solar industries

Surrounded by a dominating coal power industry, the wind and solar power industries have grown in record time over the past 15 years in China, and the country has mustered an impressive ingenuity in fine-tuning policy mechanisms to induce the growth of these new industries. To be sure, there have been considerable challenges; a quick industry buildup comes with a price. There is no doubt that the impetus behind the wind and solar industries has been more on industry creation than on the provision of renewable electricity (Zhang *et al.* 2013). This chapter tries to understand what mechanisms have allowed for the rapid growth of China's wind and solar industries in addition to the more formal laws and regulations. As established in the previous chapter, the way in which China's energy sector is governed can be described as both fragmented and centralised. In this chapter, I first briefly tell the story of how the wind and solar industries came into existence in China, before I look closer at what the various aspects of fragmentation and centralisation mean for the wind and the solar industries.

The rapid growth experienced in terms of wind turbine and solar PV installations in China represents a velocity of capacity development never before witnessed in the history of renewable energy (Mathews and Tan 2015). This development has involved the coordination of interests and an alignment of actors and institutions on a massive scale. About three-quarters of China's total solar PV installations by the end of 2018 were plugged into the grid between 2016 and 2018, implying that the growth has happened especially quickly the past years (Publicover 2019). For wind, the equivalent figure was 30 percent (Publicover 2019). State Grid was responsible for connecting and transmitting most of this electricity, and it is impressive that the amount of wasted electricity (curtailment) has been reduced at the same time as the installations kept increasing. This chapter looks behind the numbers and figures in an attempt to understand the ways in which wind and solar energy manufacturing and electricity generation has been managed in China.

The quick growth has led to many challenges in both the wind and solar industry, such as uncertainties over the long-term performance of Chinese turbines, transmission constraints for remote regions, lack of qualified personnel, delays in connecting wind and solar farms to the electrical grid, waste of electricity (curtailment) and overcapacity (e.g., Martinot 2010, Gosens and Lu 2013, Quitzow 2015; Huang *et al.* 2016; Andrews-Speed and Zhang 2019). As already alluded to in the previous chapter, fragmentation and centralisation, together with policy experimentation and policy learning, have been fundamental for policy flexibility and institutional adaptability. Building on the work of Lema and Ruby (2007), who conclude that coordination was important for the growth of China's wind industry, this chapter looks at the processes which have led to coordination, including experimentation and scaling up in China's solar and wind industries. Also covering the period from 2011 to 2018, the chapter develops a hypothesis in which coordination and fragmentation are mechanisms the government takes advantage of to control industry growth. This hypothesis will be further developed in later chapters of the book as well. These features make China's energy governance system highly flexible and adaptive, enabling and constraining growth according to policy preferences, but the same features have also been decisive for the considerable challenges the industry is facing.

Several scholars have addressed the relationship between institutions and renewable energy implementation in China. Some have provided general overviews and updates (e.g., Martinot and Li 2007; Liu 2019), some have looked specifically at the policy regime and institutions (Cherni and Kentish 2007; García 2011, 2013; Schuman and Lin 2012; Wang 2007; Zhang, Andrews-Speed and Zhao 2013), and some have looked at innovation and technology transfer (e.g., Gosens and Lu 2013; Huang *et al.* 2012; Lewis 2013; Ru *et al.* 2012; Urban, Nordensvärd and Zhou 2012; Zhou *et al.* 2012; Nahm and Steinfeld 2014). Most of these studies highlight the explosive development of China's wind and solar industries, and the considerable challenges the development has led to. This chapter looks at the institutional mechanisms on a level deeper than the formal laws in order to show how the same governing institutions have inherent advantages and disadvantages vis-à-vis the development of the wind and solar industries. Implementing legislation is a craft, and China's achievement in growing the largest wind turbine and solar PV industries in the world in the time span of little more than a decade cannot only be reduced to a set of laws. Therefore, this chapter aims to understand better what types of governing practices have laid the foundations for the rapid development.

Laws and regulations promoting wind and solar

This section provides an overview of the formal policies which have promoted China's wind and solar development, divided into two main areas: those which promote manufacturing and industry development, and those which promote a domestic market through electricity generation. Although the two areas clearly are related, it makes sense to make this division here, because the Chinese strategy to develop these industries can be seen as separate from the interest in increasing domestic consumption of renewable power (Ydersbond and Korsnes 2016). This is especially clear through the comparison of the development process of the wind and solar industries in China.

Electricity generation—wind and solar

Measures aimed at increasing the proportion of renewable electricity production in China are mainly covered in the Renewable Energy Law (RELaw), which was brought into force in January 2006, and with amendments effective April 2010. The Central Committee enacted the law with overwhelming support, suggesting that Chinese legislators almost unanimously recognised the need for renewable energy (Wang 2007). The law was drafted over a two-year period, and international and domestic experts, various types of organisations and governmental bodies provided advice and comments in order to calibrate it to fit China's ambitions (Interviews #3 and #4, Sept. 2011; Martinot and Li 2007). The law, therefore, was the result of an international learning process, in which experiences from abroad were taken into consideration before enactment. This was also the case before each of the major revisions of the RELaw in 2009 and 2010.

RELaw measures include government installation goals, mandatory market shares, a tariff system, and a cost-sharing principle through the establishment of a special fund for renewable energy development through increased power prices for residential consumers (Jiang 2011; Lewis 2013). The law states that energy authorities of the State Council shall set medium- and long-term targets for how much renewable energy should be developed nationally. Such plans were made but have since been adjusted repeatedly, because the development has happened faster than expected, making the initial targets look rather small. For instance, in plans such as the Medium- and Long-Term Development Plan (MLP) for Renewable Energy of 2007 (until 2020) or the Twelfth (2011–15) and Thirteenth (2016–20) Five-Year Plans (FYP), there were clear future installation goals. For solar PV, the 2007 MLP had a goal of 0.3 gigawatts (GW) installed by 2010 and 1.8 GW by 2020 (china.org.cn 2007). However, in 2010 the installed capacity was already 0.8 GW, and in the twelfth FYP, the target for 2015 was adjusted to 21 GW and later 35 GW (Gosens, Kåberger and Wang 2017). For wind, the target in the 2007 MLP was 5 GW for 2010 and 30 GW for 2020. Similarly, developments happened quicker than anticipated, and in the 12th FYP the target for 2015 was adjusted up to 100 GW, whilst actual installed capacity in 2015 was 130.8 GW (Gosens, Kåberger and Wang 2017).

The 'big five' government power generation companies—China Energy Investment Corporation, Datang, Huaneng, Huadian and State Power Investment Corporation—are also large investors in renewable energy since they are mandated by the government through the Eleventh Five-Year Plan for Renewable Energy to install at least 3 percent non-hydro renewable power as a portion of their total capacity in 2010, and 8 percent in 2020

(CWPO 2010). This applies to all utilities which have a capacity of more than five GW of thermal power electricity generation (CWPO 2010). These mandated market shares undoubtedly led to an increase in renewable energy investments by the large state-owned enterprises (SOEs). Zhu *et al.* (2019, p. 207) find that hearsay about these mandates was the 'primary reason' for these companies to 'expand their wind power business in the first place'. However, a downside might have been the fact that these large power utilities only cared to fulfil their installed capacity criteria and had less of an incentive to focus on the kilowatt hours (kWh) produced (CWPP 2010), something which demands more resources in terms of operation and maintenance.

In accordance with the RELaw, electric utilities are obligated to purchase all renewable power produced. An important change in this regulation appeared in the 2009 amendment of the RELaw, when the utilities were obligated to purchase all renewable power even when there is not sufficient power demand on the grid (Martinot and Li 2010). Moreover, the 2009 change pointed out that central organisation and planning of renewable energy should be followed at the local level, and 'planning at the local level should contain details of specific targets, allocations, a time line, grid connection, service systems and supporting measures' (Huo and Zhang 2012, p. 40). These clarifications likely made it more interesting for developers to also consider solar PV, and not only wind as their main renewable energy development source.

Arguably the most important measure is the feed-in tariff (FiT) established through the RELaw. These tariffs are technology specific, meaning that an electricity tariff is set for sales of electricity from the particular technology—making them competitive with other electricity sources such as coal power. In addition to the tariff, FiT policies usually involve contract durations (i.e., 20 years), a cap in total installations and a stepwise tariff price reduction (called a 'degression') mechanism, aiming to coincide with technological cost reductions (Ye, Rodrigues and Lin 2017). What is special for China is that the country has been divided into geographical tariff areas depending on the renewable energy resource in each location. Further, that there is no automatic tariff degression mechanism, and that the period of tariff adjustment is much longer than other countries (30 months) (Ye, Rodrigues and Lin 2017). For wind power, the FiT was established in 2009. Before this, wind power prices were determined in a variety of ways, and specific provinces often had their own wind power–pricing systems (Lewis 2013). For instance, Guangdong was one of the first provinces to experiment with wind technology development and set its own feed-in tariff in 2004 (Lewis 2013). This changed as of 2009, however, when the NDRC established a unified national FiT, with prices divided into four different geographical areas depending on available wind resource.

For solar PV on-grid project tariffs were from 2008 mainly decided on a project-by-project basis depending on tenders issued by the NEA, where the lowest price-bid established the FiT (Huo and Zhang 2012). FiTs were mainly used as incentives for on-grid utility-scale PV projects, and distributed and off-grid solar projects had other incentives, such as a capital subsidies and net metering (Huo and Zhang 2012). Before 2011, local and provincial governments were more supportive than the central government of PV projects by providing market incentives. Especially Jiangsu Province was at the forefront, and in 2009, they issued a plan which set a feed-in tariff as well as a planned amount of installations each year totalling at more than 0.4 GW in 2011 (Huo and Zhang 2012). Coincidentally, Jiangsu Province was also highly supportive of local PV manufacturing facilities, showing that industry and renewable electricity generation objectives were closely aligned (Huo and Zhang 2012). In 2011 a national FiT for solar PV was finally established, and in 2013 this was further revised to distinguish between centralised (or utility-scale) and distributed PV projects (rooftop- or ground-mounted PV panels close to where the electricity is consumed) (Ye, Rodrigues and Lin 2017). For utility scale, the FiT was divided into three different zones depending on the resource, and was set at 0.90, 0.95 and 1.00 *yuan*/kWh accordingly (Bin 2018).[1] Distributed solar installations were subsidised at 0.42 *yuan* per kilowatt hour, and in Chapter 7, I describe more in detail China's distributed PV experiences. Moreover, the 2013 change divided the country into three tariff regions depending on solar resource (Ye, Rodrigues and Lin 2017). The tariff period was also set to 20 years (Ye, Rodrigues and Lin 2017). The solar PV FiT was reduced several times after 2013, and on 31 May, 2018, it landed at 0.50 *yuan*/kWh, 0.60 *yuan*/kWh and 0.70 *yuan*/kWh across the three regions, with the distributed FiT at 0.32 *yuan*/kWh (Bin 2018).

Implementing feed-in tariffs is costly, and the gist of the cost-sharing policies is to evenly share the additional costs of renewable energy through a combination of civic duty and responsibility by various stakeholders. This has been achieved through the so-called Renewable Energy Development Fund, which compensates the costs from the renewable energy subsidies through an overall electricity price surcharge in China. The surcharge started out on 0.001 *yuan*/kWh in 2006, increased to 0.002 in 2008, 0.004 in 2009 and 0.008 *yuan*/kWh in 2011 (Jiang 2011). It was raised again in 2012 to 0.015 *yuan*/kWh, and since 2016 it has been kept at 0.019 *yuan*/kWh (Lin 2018). This money is to be spent on three main areas: the FiT, operation and maintenance costs and grid integration costs (Jiang 2011, p. 108). If we compare these prices to those in for instance Germany, they are relatively low. The German surcharge equals about 0.8 *yuan*/kWh, which roughly is the same as China's average end-consumer tariffs (Lin 2018). Cheap electricity is considered a highly strategic issue in China in order to ensure economic growth, and increasing power prices is not something that is done with ease. In 2017, about 5 percent of electricity bills were composed of the renewable energy surcharge, as well as other surcharges (such as the one initiated in 1993 to support the construction of the Three Gorges Dam) (Chen and Li 2019). Thus, there has been a recurring problem to cover the subsidies for renewable energy, as there are more projects accepted than income generated through the surcharge. This means that there are long delays between project initiation and when project developers receive their promised subsidies, which can generate financial problems. This problem has increased with time, and especially after the rapid increase in solar PV installations since 2015. In 2017, the NEA reported that there was a deficit of 112.7 billion *yuan* (CREO 2018). This was one of the backdrops which led to the 31 May 2018 halt in central government subsidies for solar PV (CREO 2018).

Wind industry development: experimentation and fragmentation

The modern wind turbine history of China can be divided into two phases: one which lasted until the mid-1990s, where the focus was largely on expanding the smaller, decentralised, off-grid wind turbines; and a second which focussed on manufacturing and implementing large-scale, grid-connected wind turbines (Gan 1998). The first phase was characterised by government support programmes which emphasised rural electrification, as well as the buildup of manufacturing capacities (Gan 1998). The first subsidy ever given to renewable energy in China was given from the Inner-Mongolian provincial government in 1986 for small-scale wind turbine projects (Lew 2000). By the 1990s, China already produced more small-scale wind turbines than any other country (Lew 2000). China quickly became a global leader in the manufacture and export of small wind turbines (up to 100 kW), and between 1983 and 2008 China produced a total of 508 712 small off-grid wind turbines, with an installed capacity of 57 megawatts (MW) (REN21 2009).

During the 1990s, provincial government policies shifted focus from a household level, to larger wind farms which could produce electricity for whole villages (Gan 1998). The central government started to see wind power as a future source of large-scale electricity generation, and in 1996 the 'Ride the Wind' programme was initiated to start up utility-scale grid-connected wind power (Zhao, Hu and Zuo 2009, p. 2884). This programme emphasised the introduction of foreign technologies, as well as supporting local manufacturing of turbines (Zhao, Hu and Zuo 2009). This dual path of both building a domestic manufacturing industry as well as ensuring actual electricity generation from grid-connected wind turbines from the outset has been a core strategy for the Chinese government (BNEF 2010). Most of the technology employed in the beginning was supplied from foreign wind companies from Denmark, the United States, the Netherlands or Germany (Gan 1998). This was despite the fact that wind turbine prototypes ranging from 100 kW to 200 kW were tested and demonstrated by local manufacturing firms between 1985 and 1995 (Gan 1998). The failure in commercialising these Chinese designs resulted in a reliance on foreign turbines, which were cheaper at that time. In 1996, only 1.7 percent of the 56 MW of installed wind capacity in China was produced by Chinese turbines (Gan 1998). Large foreign companies, such as Vestas and BONUS (today Siemens) became increasingly interested in the Chinese wind market at the end of the 1980s (Gan 1998), and until 2010 Vestas still ranked amongst the top-ten-largest wind turbine providers in China in terms of market share (CWPO 2010).

Even though the direct indications are difficult to deduce, it is highly plausible that the focus on small-scale wind power made it easier for the government to promote a legislative framework for grid-connected turbines. The development of the larger wind turbines started in the 1990s and grew quickly after 2003. Yet, it appears that not much spillover happened between the small-scale and large-scale types of wind turbines in terms of technology innovation. Instead, technology was either licensed from abroad or jointly designed with a foreign company, something which became the prime model of wind turbine development in China (Lewis 2011, p. 287). Today, several Chinese manufacturers possess the key manufacturing technology for MW-scale wind turbines and are capable of doing independent research and development (R&D). Also, the industrial manufacturing chain for wind turbines is relatively complete domestically, although manufacturing capabilities for some key parts, including high-precision bearings, control systems and converters, to some extent depend on imports (CNRED 2011).

As an example, Goldwind is the oldest large-scale Chinese wind turbine manufacturer, established in 1998 (Lewis 2007). This company started out as a wind farm developer in Xinjiang Autonomous Region, the western-most province in China (Lewis 2007). Goldwind licensed their first turbine from a small German turbine designer, Jacobs, which was later acquired by the German manufacturer REpower (Lewis 2007). Goldwind has since licensed turbine designs from REpower and Vensys—another German manufacturer—which designs direct drive turbines (turbines which do not have a gearbox). In 2008, Goldwind became the majority owner of Vensys (Lewis 2011). Some of the licensing agreements have certain requirements, such as a prohibition of usage outside of China, or limitations on future design modifications, as, for instance, for the REpower licensed turbine for Goldwind (Lewis 2011). This also applies to other Chinese manufacturers. For example, Mingyang, one of the largest Chinese turbine manufacturers, is only allowed to manufacture and distribute their aerodyn-licensed turbines domestically (Mingyang 2011). Such restrictions may also have had impacts on the limited degree to which Chinese wind turbines have been exported, as we will return to in Chapter 5.

Furthermore, the licenses were not cheap. As an illustration, the licensing agreement Goldwind had with Jacobs required the company to pay a EUR5,000 royalty per turbine produced (Lewis 2007), and Mingyang currently has to pay a royalty between 0.5 percent and 2 percent of the sales price, depending on the number of turbines sold (Mingyang 2011). In 2009, an estimated US$450 million, covering patent royalties, licenses and technical service, flowed from China to Europe and the United States, financing research and development of wind turbines there (CWPO 2010). All in all, the strategy the Chinese industry chose for technology acquisition is well reflected in the slogan of the Chinese-Dutch wind turbine producer

XEMC-Darwind: 'Dutch design integrity & Chinese industrialization power' (XEMC-Darwind 2012). In the next chapter, we will look closer at the various technological learning mechanisms the Chinese wind industry employed. We now focus more on the factors which facilitated the rapid growth.

At least three important factors directly promoted the rapid development of China's wind industry. First, the domestic content requirement of wind turbine manufacturing in China led to the development of supply chain markets; second, speedy approval for wind power projects at a provincial level resulted in a huge increase in the number of wind turbines each year; and third, relatively unproblematic access to finances. The last factor will be discussed more in detail later, but suffice to say here that in 2011, China attracted US$52 billion in new renewable energy investments, 60 percent of which went to wind projects (FS-UNEP 2012). Since 2011, China has attracted the most financial investments for new renewable energy in the world, with an increasing distance to number two on the list—the United States (Lin 2018). Also part of the early financing and an important contributory factor to the rapid growth until 2012 was the access the Clean Development Mechanism (CDM) through the Kyoto Protocol, which gave projects access to additional finances. In fact, until 2012, when the Kyoto Protocol ended, '64GW or 80% of China's total installed turbine capacity had obtained CDM registration' (Hayashi, Huenteler and Lewis 2018).

In 2002, the Chinese government decided to stimulate the development of wind energy through a national wind concession programme, allocating selected sites for wind farm construction to the company bidding the lowest electricity tariff (CWPP 2010). Some prerequisites were made in order for projects to be accepted, such as restrictions on turbine size and local content. In effect, the price of electricity not only decided who won the bid, but also the extent to which the turbines were manufactured locally (Wang 2010). Because of a dependence on expensive, imported turbines, the Chinese government decided that a domestic content requirement pertaining to wind turbines was needed to facilitate domestic manufacturing of turbines and turbine parts (Howell *et al.* 2010). During the first concession round, which started in 2003, the local content requirement of turbines was set at 50 percent; in 2004 this share was increased to 70 percent, and in 2009 it was finally phased out (Wang 2010). In addition to content requirements, import tariffs on preassembled wind turbines were at 17 percent in 2007, whilst tariffs on their components were set to only 3 percent (Martinot and Li 2007). This policy, together with the removal of local content requirements in 2009, is thought to have 'allow[ed] domestic manufacturers to more easily access wind components from foreign suppliers as they buil[t] the prototypes for their larger turbines' (BNEF 2010). Thus, the local content policy was central to the inauguration of the domestic wind turbine manufacturing industry.

A central point made by Lema and Ruby (2007) is that the period prior to the national wind concession programme, which began in 2002, was dominated by an extensive fragmentation of authority—for instance, in deciding whether to establish a domestic industry or rely on imported turbines. With the concession strategy, however, this fragmentation changed, and the National Development and Reform Commission (NDRC) took a more active role in coordinating the supply of and demand for wind power. Lema and Ruby also note that coordination between the trade and industry departments is what sparked the domestic wind turbine industry. This change in the status of the NDRC was undoubtedly important; however, the concession projects were arguably also an experimental point for policy development. Between 2003 and 2007, there were five concession rounds totalling 2.6 GW of installed wind power capacity, against a total of 6 GW of installed wind power at the end of 2007 (Jiang *et al.* 2011), amounting to 43 percent. Each concession round grew in size, starting at 200 MW and ending at 950 MW. Between each of these rounds, policy was changed and refined. For instance, in order to prevent developers from bidding at unacceptably low prices to secure the right to develop a wind farm, the criterion changed in 2005 from 'lowest-price bid wins' to various factors outside of price being weighted. Initially, the price was weighted at 40 percent; it was further reduced to 25 percent in 2006 (Li *et al.* 2007). Other criteria became more important for winning a bid: domestic manufacturing content, overall capability, technical planning, grid price and economic benefit, each having been given a weighted score (Li *et al.* 2006). The concession rounds provided important lessons in shaping the pricing mechanisms of the Renewable Energy Law, where 'government-guided' prices were decided on the basis of the concession project pricing (Martinot 2010). These prices were, in turn, at the base of the nationwide FiT prices (implemented in August 2009), and determined prices for four different geographical zones sorted by wind-resource quality (Martinot and Li 2010; Wang, Qin and Lewis 2012). As the interviewee from a wind energy advisory organisation (Interview #1, Sept. 2011) pointed out:

> The concessions are only a small share of the whole wind development. The government is using [the concessions] as small projects that demonstrate what the government want the wind industry to be; they want it to be modernised, to be bigger, to rise and lead [...] the global trend.

What started as an experimental policy in 2003 was scaled up and laid the basis for both policy learning and further refinements in wind power development up until 2009. The NDRC's coordinating role in the concession projects was important, but the concession rounds in themselves were useful for experimenting and gaining experience with pricing policies, which facilitated the coordination of further wind power projects through the national FiT.

The government initiated several programmes to induce wind power growth. One important planned programme is often referred to as the 'Three Gorges of the Air', consisting of seven large wind power bases of more than ten GW each, which was started in 2008 (CWPO 2010, p. 7). This programme covers a major part of the added wind power capacity, and a planned capacity of 138 GW will be produced from these bases in 2020, provided the grid capacity is adequate (CWPO 2010). The rapid development of the wind industry has led to several technical challenges which needed to be addressed. The main challenges were related to the performance and quality of Chinese turbines, with uncertainties over their long-term performance, transmission constraints for remote regions, a lack of qualified personnel, and time lags in connecting wind farms to the electricity grid (Martinot 2010).

Until 2011, China's wind industry experienced almost exponential growth, but after 2011 the average growth rate stabilised. This stabilisation is relevant for the governance of the wind sector; as coordination premised its rapid development in 2003, it was also coordination which led the expansion to a halt in 2011, by centralising the approval of new wind farms. In fact, between 2003 and 2011, more than 90 percent of constructed wind farms in China were approved by local governments—something which led to a mismatch between local wind farms and centrally planned power grid construction (Li *et al.* 2012). These local governments handled each tender application efficiently, and new projects were rolled out quickly. Included in the aforementioned concession rounds were projects of more than 50 MW, which needed approval from the central government (administered by the NDRC). Projects below 50 MW could typically be approved by local governments, and this led to large numbers of projects sized at 49.5 MW, many of which were installed right next to each other, making their real sizes much larger (Jiang 2011). As of 2011, China centralised this decision, and all wind projects were then required to be approved by the National Energy Administration. This new legislation approved a total of approximately 27 GW for the Twelfth Five-Year Period (up to 2015), 13 GW for state-approved projects and 14 GW for those locally approved (Li *et al.* 2012). In April 2012, a second group of approved projects totalled 15 GW. Any projects which were not approved through this bill were not accepted (Li *et al.* 2012). In addition to 18 new technical standards issued in 2011—limiting access to turbine manufacturers which did not adhere to these standards—the change in the approval process considerably altered the growth of new wind farms in China. Thus, decentralisation of authority was beneficial for the speedy growth of China's wind industry, and, by centralising this authority, growth slowed. In other words, fragmentation and centralisation were characteristics the government could draw on to reach development targets.

The reason local governments decided to approve projects so quickly can be attributed to their quest for economic growth (Naughton 2010, 2017). What we observe here is a divergence between provincial and central government interests which characterise Chinese politics. Local governments were increasingly concerned with stimulating local economic growth, and wind power projects were attractive to local economies looking for a boost (Interview #7, Oct. 2011). Although provinces have become more economically independent from the centre (Saich 2011, p. 183), central government approval also shapes provinces' opinions on profitable investments. As a result, when a company or sector receives central support, they are considered a safer bet for provincial governments seeking to build up an industry.

This makes a difference when local governments face the choice of whether to start a wind project or a thermal power project.

Solar industry development: private entrepreneurs and local support

Development of China's solar industry has a quite different story than the wind industry, because the solar PV industry received practically no support from the central government in the early 2000s. This marks an important and interesting difference to the way in which the wind industry emerged. The solar PV industry was driven by a handful of entrepreneurial individuals who started private companies and (eventually) received support from international investors, and the industry targeted international markets to sell PV panels on instead of domestic markets (Binz and Anadon 2018; Huang *et al.* 2016). Around the year 2000 no one expected that the Chinese solar PV industry would become world leading in less than ten years (Binz and Anadon 2018). Before 2010 on average 95 percent of PV products in China were exported mainly to markets in Germany, Italy and Spain (RED 2011; Huang *et al.* 2016). In 2011, China captured about 50 percent of the world manufacturing capacity of PV solar cells (excluding Taiwan) (REN21 2011), up from about 1 percent in 2000, and 5 percent in 2004 (Binz and Anadon 2018; Marigo 2007). Between 2000 and 2012, there was a very strong correlation between European installed PV capacity, and China's PV cell production output (Huang *et al.* 2016). China's domestic solar PV installations started to take off around 2009, when the Renewable Energy Law had clarified several of the risks which were involved with investments (Huo and Zhang 2012). This coincided with the international financial crisis and cutbacks in PV policy support in the target markets, and the collapse of the Spanish market in 2009, severely impacting the PV exports from China (REN21 2009; Quitzow 2015; Huang *et al.* 2016). Moreover, the PV exports were further affected by anti-dumping investigations by the United States and the European Union starting in 2011 (Quitzow, Huenteler and Asmussen 2017). After a feed-in tariff on PV was installed in 2011, domestic installation of PV panels skyrocketed, and grew with over 300 percent in 2010 and more than 500 percent in 2011 (Quitzow 2015).

PV cell production lines were introduced to China in the 1980s, mainly coming from Canada and the United States, and by 1995 there were six solar cell–manufacturing companies in China (Dai and Shi 1999). From the end of the 1990s and the beginning of the 2000s, new solar cell and module producers emerged. Local and provincial governments did provide some support in shape of refunding land transfer fees, corporate income tax and cheap electricity, hoping to boost local GDP (Zhang and Gallagher 2016). However, much of this support came in the shape of general high-tech support schemes not particularly interested in attracting PV industries (Binz and Anadon 2018). Therefore, in the very early days of the solar PV–manufacturing industry (1997–2004), the few handful of companies which made it, the first wave of start-ups, such as Yingli (founded 1997), Suntech (founded 2001), Trina (founded 1998) and Canadian Solar (founded 2001), had to actively work to convince local governments about the benefits of the PV-manufacturing facilities (Binz and Anadon 2018). The founder of Suntech, for instance, pointed out that he had to make 'proposals to a number of local governments across the country before finding a receptive audience in Wuxi' (Zhang and White 2016, p. 607). Only after 2007, when the PV industry had become successful internationally, did local governments start supporting the Chinese PV industry through subsidies and incentives (Zhang and White 2016). Some of the initial PV companies used independent technology, but most key production line equipment was imported from Europe and the United States (Marigo 2007, Zhang and Gallagher 2016). Key components and machinery were largely imported and adapted with domestic materials such as cover glass and cables to reduce costs but still keep high quality (Binz and Anadon 2018). The first wave of solar PV start-ups did not have the availability of highly automated turnkey manufacturing facilities, which only entered the market in 2006 (Quitzow 2015), and therefore assembled own manufacturing lines or imported secondhand equipment from other countries (Binz and Anadon 2018).

The Chinese solar PV industry picked up rather quickly during the early 2000s, and many of the companies were started by Chinese who were educated abroad and came back to found or join new companies (Gallagher 2014). The second wave of Chinese solar PV start-ups came around from 2004 to 2006, with companies such as Sunergy, Jinko Solar, LDK Solar and JA Solar (Binz and Anadon 2018). These companies benefitted from a more automated equipment– manufacturing industry with turnkey facilities, that is, that you practically only need to 'turn the key' and the production starts, that could be imported and set up quickly without much previous skills (Huang *et al.* 2016). This 'plug and play' feature of the solar PV industry allowed the industry to move in and scale up quickly (Huang *et al.* 2016). After 2005, provincial and local governments started eyeing these new solar firms as an opportunity to create local jobs and perhaps boost local economic growth (Gallagher 2014; Zhang and White 2016). As an example, the city of Huaian in Jiangsu Province

> provided a 50 percent refund in the real interest of loans used to purchase equipment for a factory, a refund equal to 0.05 RMB/kilowatt-hour (kWh) in electricity consumption for the first year, a refund of the land transfer fee, and a partial refund of corporate income tax for the first eight years.
>
> (Gallagher 2014, Kindle loc. 1452)

Chinese companies started out mainly producing solar cells and later on expanded into wafers/ingots, PV modules and finally silicon production, moving from less to more technologically advanced capabilities, and several companies aimed to cover a large part of the supply chain to become wholly integrated manufacturers (Shubbak 2019). Marigo, Foxon and Pearson (2010) and de la Tour, Glachant and Ménière (2011) indicate that managerial competences and the presence of internationally trained managers were advantageous for the Chinese solar industry. These highly skilled executives brought 'capital, professional networks, and technology acquired in foreign companies or universities to China', and more than 60 percent of the board members of the three-largest PV manufacturers had experience from abroad (de la Tour, Glachant and Ménière 2011, p. 765). Compared to the wind industry, licensing of technology was not an important way of acquiring PV technology competencies (de la Tour, Glachant and Ménière 2011). Partnerships with foreign supplier companies, labour mobility and to a lesser degree foreign direct investment were important avenues of acquiring and mastering PV technology (de la Tour, Glachant and Ménière 2011).

Highly purified silicon is one of the core materials in crystalline PV panels, and the provision of this material became a bottleneck between 2004 and 2008 due to the rapid increase in PV demand (Zhang and Gallagher 2016). The shortage of polysilicon led to increases in prices, and this was a motivating factor for some Chinese companies to attempt to produce their own silicon (Zhang and Gallagher 2016). Already in 2002, the Chinese government started directed efforts at mastering the purification process of silicon, as there were just six companies globally, from the United States, Germany, Norway, Japan and South Korea, who dominated the market (de la Tour, Glachant and Ménière 2011; Shubbak 2019). In 2006, China had 37 percent of patented innovations in silicon production, although only 2.5 percent of the silicon was produced in China (de la Tour, Glachant and Ménière 2011). This R&D effort was mainly supported by government funding, with a strategic aim to 'break a technological lock-in in a strategic segment' (de la Tour, Glachant and Ménière). Thus, the Chinese government must have had some strategic ambition behind the support of silicon R&D, knowing that this could become a potential bottleneck. And indeed, it appeared like the Chinese industry succeeded with the directed effort. Since 2010, five Chinese companies were able to catch-up and expand their production capacity, and in 2015 they occupied 38 percent of the global market share in silicon production (Shubbak 2019). In Chapter 5, we will return to the highly interdependent relationship between China and international actors and events in the PV industry.

Since Chinese solar cell manufacturers exported such a high quantity of cells, they were exposed to any global imbalances and trade restrictions. As incentive mechanisms for solar energy production in Germany, Italy and Britain were being reduced around 2011, sales decreased, and the Chinese solar industry experienced rough times (reuters.com 2011). Only after these companies had entered the world market and contributed to considerably lower the price on PV panels did the Chinese central government enter with new policies for domestic market formation (Gallagher 2014), as already described in the section about FiTs. It should be noted, furthermore, that as the domestic installation of PV panels started growing more rapidly around

2013, this was also related to the fact that local governments could approve projects locally (Bin 2018). This led to a rapid growth in not only centralised utility-scale project, but also distributed PV projects, testifying to the fragmented nature of energy governance also in the PV industry. For example, although the central government had established PV installation quotas for each province in 2014, the five northwestern provinces Shaanxi, Gansu, Ningxia, Qinghai and Xinjiang installed more than twice that quota in 2015 (Ye, Rodrigues and Lin 2017). These are also provinces which struggled with high shares of renewable energy curtailment (Ye, Rodrigues and Lin 2017). (Ye, Rodrigues and Lin 2017). Moreover, Xinjiang and Gansu were not allocated a quota at all in 2016 and yet 'there were about 3.3 GW and 0.7 GW installed in Xinjiang and Gansu respectively' (Ye, Rodrigues and Lin 2017, p. 500). In 2017, an astounding 53 GW in solar installations were added, of which 20 GW was distributed PV projects (rooftop and ground mounted). In 2018, there was 44 GW of newly installed capacity, of which 48 percent was distributed (chinaenergyportal.org 2019a). This implies a rapid increase in the amount of distributed solar PV projects, to which I will return in Chapter 7.

Coordination, legitimacy and centralisation

In the previous section, we presented some mechanisms which promoted industry development and renewable electricity markets. Now we go into a bit more detail to get a better understanding of what was of relevance beyond the more formal support mechanisms. In order for a new industry to come into existence, a certain level of public acceptance and legitimacy is required—and such legitimacy is often established through actors which already have legitimacy, such as governments or other, related industries. In China, important policy measures to increase the share of renewable energy have been directed at established energy companies, which already have built up solid legitimacy in a related industry. An example of this is the government's introduction of the mandated market share of non-hydro renewable energy for the established (coal) power producers in China. This mandate was a clear signal that the road to renewables was to go through the existing power utilities, irrespective of their ties to coal and hydropower. Another strong signal of commitment is provided when areas of priority are decided through long-term plans by central and local governments. The Five-Year Plan is the most important government document, and the attention given to new and renewable energy has increased over the course of 13 Five-Year Plans, beginning with the sixth and culminating with the thirteenth plan covering the period of 2016 to 2020 (Yuan and Zuo 2011; Gosens, Kåberger and Wang 2017). Without a doubt, these government indications play an important role in paving the way for emerging industries. This section explores further sources of legitimacy which were important in the development of the wind and solar PV industries.

Control of media

In the Chinese wind turbine industry, the role which the media could play was particularly observed during the 2011 downturn. Several informants mentioned that during 2011 stories about quality issues and poorly functioning turbines started appearing in the media. As the Chinese media is largely state-controlled, the Chinese Communist Party (CCP) can dictate what is allowed to report. The interviewee from a wind energy advisory organisation highlighted this (Interview #1, Sept. 2011):

> [The government] can one day say that "we think this industry is very promising" and everything [is] good about it; [...] everything you can see related to wind in the news is good. This reinforces the industry to expand. Now, this year, it's a time when some of the problems that were hidden started to get exposed. [...] Every problem was there two years ago; it's just that people were not allowed to say it, so it didn't seem to be there. But now we are suddenly allowed to say that [there are problems], and people get a feeling that wind started to show its side effects. But that's not true, the side effects have always been there; it's just that they are exposed at this stage.

This quote illuminates the importance of the media in establishing the legitimacy of the wind industry. What we observe is a government that uses legitimacy as a tool to increase or decrease interest in the wind industry, fluctuating with current development goals. To be sure, highlighting the challenges which the industry faces regarding turbine quality or grid connection issues is important for the overall performance of the industry. Yet, the fact that these issues were evident for some years prior to 2011 and there had not been any repercussions testifies to the importance of information control in China. This form of (de)legitimation can be said to be integrated in the CCP policy toolbox, and its usage ultimately depends on the degree to which the government (the CCP) and industry goals are in sync. The control of media could therefore both have contributed to rapid growth in the period until 2011, and when a focus on quality finally emerged it contributed to slowing growth.

Politics and legitimacy through related industries

The Chinese wind industry gained legitimacy within established institutions by using the preexisting influence of energy incumbents to shape outcomes. Many of the large, influential state-owned enterprises (SOEs) engaged in wind turbine manufacturing, and the regional political influence of the SOEs facilitated their growth. According to the informant (Interview #6, Oct. 2011) from a wind turbine manufacturer, which is a subsidiary of a large multi-industry conglomerate, the company 'has a certain influence in Hunan Province, because the governor of Hunan Province came from (this company)'. As a result of the company's political connections, it has been able to convince policymakers of the benefits of wind turbines. Indeed, there is a well-documented link between SOEs, economic performance and political careers (Andrews-Speed 2011; Li *et al.* 2008; Xu 2011). This relates to the role of the Ministry of Personnel, which has the capacity to appoint or dismiss the senior executive leadership of even large SOEs, as already elaborated on in Chapter 2. Often, industry professionals are appointed to these positions because of their technical insights, and these positions are, in turn, used as stepping stones for political careers, similar to that of the XEMC executive (Rosen and Houser 2007). This means that the leadership of large energy companies must be attentive to party politics, and balance central political demands against personal ambitions and provincial needs. All large, state-owned energy companies are mandated (through the Renewable Energy Law) to produce electricity from renewable energy sources, and failure to comply can hamper company advancement and political careers. One interviewee (Interview #2, Sept. 2011) put it this way:

> The incentive for the leaders of the power companies to fulfil this requirement is that they will be held accountable if targets are not reached [...] and this will directly impact their personal careers.

Moreover, central government support can mean more in terms of politics than in terms of economic benefits they provide. This was demonstrated in several interviews when talking about the role of support for research and development (R&D)—for instance, with one large wind turbine manufacturer (Interview #6, Oct. 2011):

> The company's own investment is larger than government funding, but state funding is also very important to us; it shows that the state encourages [us] to keep up. Especially for our group, a very large SOE with a long history, the state funds mean more [in terms of] encouragement than [...] real impact. After the state funding, we have more voice in Hunan Province, which means the Hunan provincial government would be more supportive to us.

A government official from a government policy-and-advisory agency presented a similar line of argumentation (Interview #8, Nov. 2011):

> [The government] has promoted R&D a bit; some national research centres and test centres have been supported by the government. And that has been enough because it proved the legitimacy of the industry.

Central government subsidies and support eliminated some of the risk local governments faced in choosing their investment strategies, and it allowed large companies to be more confident in entering a new industry such as the wind industry. The central government, therefore, incentivised SOEs to follow its legislation by appealing to a company's success and the benefits to the political careers of managers, and it also made a company's success more likely by showing its support.

The political power and legitimacy accumulated in other industries also benefitted the wind industry. Most of the large turbine manufacturers in China have parent companies from related industries within machinery and equipment manufacturing, as well as direct links with electric power utilities. A case of the latter is Guodian United Power, a subsidiary of Guodian, which is one of the five state-owned electric power companies and China's largest wind power installer (Li *et al.* 2012). Since Guodian United Power was established in 2007, it grew to become the third-largest Chinese turbine manufacturer (a milestone reached in 2013) and it was one of the fastest-growing companies in the wind industry (WWEA 2014; Li *et al.* 2012). The company benefitted greatly from the unique position of its parent company in wind farm development, which was also emphasised in an interview with the deputy director of United Power's Chief Engineering Office, Mr. Xiao Jinsong:

> United Power has the advantage of control throughout the entire supply chain. In addition to providing the complete machine, we also produce major components—blades, gearboxes, generators, pitch systems, inverters, etc. Furthermore, our parent company, Guodian, is the largest wind power developer in Asia.
>
> (DNV 2011)

The experience of large industrial companies was central for the advancement of many wind turbine manufacturers, many of which have come from the coal power equipment manufacturing industry. The three largest coal power equipment manufacturers, Shanghai Electric Group, Harbin Electric Corporation and Dongfang Electric Corporation, which provided nearly all the advanced coal power equipment in China (Yue 2012), all had subsidiaries in wind turbine manufacturing. For instance, the central government–administered Dongfang Electric, the fourth-largest Chinese wind turbine manufacturer in 2013 (WWEA 2014), has a history of more than 50 years in manufacturing heavy-duty machinery and equipment, such as steam and hydro-turbine generators (DEC 2012). This company, one of the largest steam turbine producers in China, did not engage with the wind industry until 2005, when it started cooperating with European turbine design companies (Zhao, Hu and Zuo 2009). In addition, Shanghai Electric signed strategic alliances with the European firm Siemens. All these companies became large actors within the Chinese wind industry by exploiting their established legitimacy within China's institutional framework.

In the solar PV industry, the story is again somewhat different. Legitimacy there was initially not sourced from the CCP, but through individual entrepreneurs which were able to convince local governments about their solar PV manufacturing plans (Zhang and White 2016). What is more, these first movers were also able to convince international markets that products of good quality could originate from China, which was not common belief in the early 2000s. The second wave of entrants into the Chinses solar PV manufacturing industry from around 2005 were able to draw on this legitimacy which was already established by the first movers (Zhang and White 2016). In other words, the first movers in this case were able to overcome a 'liability of newness', which also benefitted second movers, since key resource holders, such as suppliers, managers, customers, investors and local governments started seeing the PV industry as promising (Zhang and White 2016). Another important source of legitimacy came from German market actors, where there was a mismatch between supply and demand of PV panels, creating a window of opportunity for Chinese PV manufacturers (Quitzow 2015; Zhang and White 2016). German PV firms entered into various agreements with Chinese PV firms, and the credibility that these German firms already had in the German market 'played a crucial role in overcoming initial customer hesitation and establishing China's PV manufacturers as legitimate competitors' (Quitzow 2015, p. 135).

State banks and preference for SOEs

One of the drawbacks of the central government's strategic usage of SOEs and established companies is related to the preference of industry creation, and hence quantity before quality—often leading to overcapacity. Overcapacity issues have been seen in both the wind and solar PV industries. In the wind industry it was reported in 2009 that more than 70 different wind turbine manufacturers were attempting to get a piece of the market (REN21 2009). By 2012, production capacity far exceeded domestic demand, and more than 40 percent of China's wind turbine equipment manufacturing industry was standing idle (Andrews-Speed and Zhang 2019). In the solar PV industry, the overcapacity at the time was even worse. Around 2012, it was reported that more than 2,000 companies in 300 different Chinese cities were involved in the solar PV–manufacturing industries, and Chinese manufacturers produced more than two times the global demand of solar PV panels in 2012 (Andrews-Speed and Zhang 2019, p. 148). In short, local governments and companies were overly enthusiastic about these new industries. The central government mustered several ways to attempt to relieve the situation. One was, as already mentioned, to start supporting the domestic installation of PV panels. Another way was to force consolidation, as was the case with Suntech in 2013, when the company went bankrupt and was taken over by Chinese investors backed by local government banks (reuters.com 2013).

Indeed, an important regulatory mechanism which influenced both the wind and solar PV industries was the government-controlled Chinese banking system. In 2009, four major commercial banks accounted for more than 70 percent of China's financial assets (Walter and Howie 2011). The main task for Chinese banks has largely been to support the SOEs, even after economic reforms of the banking and finance sectors (Walter and Howie 2011). According to Saich (2011), commercial banks are directed to lend to SOEs, even though three so-called policy banks, which specifically look after government-mandated lending, were created in 1994. In effect, this means that the banks are instruments which the government can use to achieve development goals. One key measure the government used to control the market is the deposit reserve ratio (Interview #6, Oct. 2011): the minimum reserve amount each bank must hold of customer deposits. This ratio was increased several times before 2011, making it more difficult for the banks to lend out money, contributing to the 2011 slowdown of the wind industry, and the consolidations happening in the solar PV industry.

In order to fully grasp what this meant for the wind industry, recall that the largest investors in wind power projects were large energy investment companies owned by the central government. These SOEs obtained loans more easily than private competitors, and this preference for SOEs could, in turn, cause scale advantages, as pointed out by the interviewee from a wind energy advisory organisation (Interview #1, Sep. 2011). 'In general [...] [the SOEs] will order in large quantities and of course this will give them an advantage in negotiating the component's price, and that of course will get the price down'. Even though the support of large SOEs created rapid growth, their preference could lead to a lack of project evaluation behind credit decisions. This concern was expressed by an experienced wind energy consultant in China (Interview #10, Nov. 2011), who commented:

> The government approves the projects and the money is going to state-owned companies, so within the state sector there is no major perception of risk. On the one hand, there is no technical and commercial diligence, or specifically what we would call 'project finance' behind most wind farms. But on the other hand, when the signal is alright, every loan officer or every bank knows it's not a bad idea to lend money to wind projects, and that has enabled the wind industry to flourish.

In other words, the preference for low-quality, state-owned projects, induced by government investment, is risky and potentially destructive—depending, for instance, on the relationship between the supplier and manufacturer in question (Korsnes 2016).[2] In the next chapter, I return to this very different way of evaluating risks in Chinese industry development strategies.

Discussion and conclusions

There are several similarities between the way in which China's solar PV and wind industries emerged. Although the solar PV industry relied on an international market initially, the continued growth of the industry was very much dependent on central government market-formation policies after 2009 (Quitzow 2015). In terms of manufacturing, different types of local government support mechanisms were important for both the solar PV and wind industries in the early days of each industry (Nahm 2017). Chinese firms in both the solar PV and wind turbines industries did not seek to follow central government policies on emulating foreign technologies and developing local capabilities, and neither did they uniquely follow local government incentives for mass manufacturing. Rather, Nahm (2017, p. 723) points out that there was a beneficial synergy in the divergence between central and local policies: 'local manufacturing policies have complemented national R&D programmes and expanded the types of resources available to wind and solar firms'. What is more, in the learning processes chosen by Chinese solar PV and wind firms, there were several global collaborative efforts which not only benefitted local Chinese firms, but also international learning processes, leading to 'better, faster and cheaper' commercialisation of new renewable energy products globally (Nahm and Steinfeld 2014, p. 298).

Going more into detail about the particular types of knowledge which were learned in the solar PV and wind energy industries in China, Zhou, Pan and Urban (2018) find that there are differences in the global knowledge links in the two industries, and that this difference relates to the fact that these two types of technologies are relatively different:

> By acquiring manufacturing equipment, transfer of explicit knowledge, and mobility of skilled workers across borders, Chinese PV firms have built strong links through inter-firm collaboration across the world, while Chinese wind firms have weak links focusing on joint development or protocol, relying on a high degree of product-specific tacit knowledge.
>
> (Zhou, Pan and Urban 2018, p. 28)

Based on this, Zhou, Pan and Urban (2018) conclude that the solar PV industry in China is more innovative than the wind industry, and that this was related to the fact that the solar PV industry had to be internationally oriented from the beginning. Quitzow, Huenteler and Asmussen (2017), to a large degree, support these conclusions, pointing out that technology related factors and the different modes of international technology transfer were the main reasons behind the differences between the development trajectories of the solar PV and wind industry. However, seen from the point of view of China's wind industry, the CTO of Mingyang at the 2018 Offshore Wind China Summit, wind turbines were more complicated than solar PV:

> A PV panel is very simple. If we think about this in terms of species, then wind turbines are the more complex species. Wind turbines today are so complex with so many different variables to control. This forces us to do innovation. It's not like 1 + 1 = 2. That's a simple problem. Using robotics and machine learning, a turbine can learn and become better, and develop as a species. We need to treat turbines as a species and not a cold technology. It evolves, and only the fittest will survive.

Thus, although China may have been more innovative in PV technology, this could also be related to the fact that wind turbine technology is more complex.

In another recent study, Binz *et al.* (2017) compare the development of the wind, solar PV and biomass industries in China, and conclude two main things: One, the industries were highly interdependent on international conjunctures which related to knowledge, markets, investments and legitimation dynamics. Two, 'policy measures were continually and skilfully evaluated and revised based on feedback from the industry and intermediary actors' (Binz *et al.* 2017, p. 433). Thus, the ways in which Chinese policymakers adapted policies to the development of both industries played important roles.

We cannot say that the choices of renewable energy policy incentives which China chose were particularly innovative. In fact, successful renewable energy promotion policies have tended to be very similar globally (Lewis and Wiser 2007). But the way policy has been implemented—by combining experiments, which are then scaled up, with a fluctuation of central government involvement in the solar and wind industries—has proved effective. Fragmented authority in China's policymaking system has been conducive to implementing renewable power sources, as shown through the burgeoning manufacturing industries, and in the example of speedy approvals of wind farms until 2011, and of solar PV projects after 2013. Local governments conducted these approvals, and when misalignments occurred between central and provincial government development goals, centralisation of decision-making slowed development. At the same time, the experimental basis of policy development has gradually developed a larger framework for domestic and foreign wind industry actors in China. A similar experience can be said to have been the case with distributed solar PV in China, to which I return in Chapter 7. Thus, this alternation between centralisation and fragmentation has demonstrated that the orchestration between the two sides of the pendulum can be said to have benefits and disadvantages, but it achieves its purpose: the promotion of the renewable energy industry.

The wind and solar energy industries are both to some extent in place because of the Renewable Energy Law (RELaw). However, this law would not have been very useful if there were no enforcement mechanisms in place. Still, several aspects of the law do not have enforcement mechanisms in place (Liu 2019). Moreover, the law lacks clear formulation of responsibilities (Li *et al.* 2012). These two problems together reduce the commitment of grid companies to acquire wind- and solar-generated electricity. For instance, the law requires grid companies to acquire all electricity they produce from renewable energy, but the wording 'guaranteed acquisition' is not adequately defined. This leaves room for interpretation, and grid companies end up curtailing (wasting) wind and solar power without any repercussions (Li *et al.* 2012). Liu (2019) further points out several other issues with the RELaw, such as the complexity of China's policy system, lack of sufficient subsidies and industry and market monopoly. This latter point relates to the fact that fossil fuel–based energy industries 'are the major tax-payers in local places and may even be invited by local government to invest under preferential policies' (Liu 2019, p. 217). Local government may also want to protect local industries, and this can create problems for renewable energies, whose electricity will anyways be exported and not used locally (Liu 2019).

As mentioned in Chapter 1, a core issue is the distance between electricity demand centres and renewable energy resource and power generation areas in China. Transmitting electricity over distances from 1,500 to 2,500 kilometres requires a high-voltage transmission grid, which is expensive and challenging to build. Until 2012, China's investments in power grid development as a portion of total investments in the electricity sector were low in comparison with developed countries (Li *et al.* 2012). This was also a result of very high investments in generation capacity, and it led to an unbalanced development of electricity generation and transmission infrastructure, and this in turn reduced the generation capacity (Li *et al.* 2012). Nevertheless, as Xi Jinping came to power, the investments in electricity transmission, and particularly ultra-high voltage transmission capable of transmitting electricity over great distances was ramped up (Yi-chong 2017). This long-distance grid was seen as an important piece in the ambition to reduce air pollution in large cities, and between 2011 and 2015, State Grid intensified investments in grid infrastructure from 302 billion *yuan* to 452 billion *yuan* (Yi-chong 2017). What is more, since 2012, the central government and State Grid have greatly improved the curtailment rate of wind and solar PV electricity generation. At the end of 2018, the curtailment rate of wind was only 7 percent, compared to percentages between 12 percent to 16 percent from the preceding five years, and the curtailment rate of solar PV was only 3 percent at the end of 2018, compared to around 10 percent during the years before (CREO 2018, chinaenergyportal.org 2019a; chinaenergyportal.org 2019b).

García (2011, p. 8048) argues that gradualism, consisting of experimental and incremental policymaking, creates barriers in China such as 'legal insecurity, fragmentation of bureaucracy[, and] targets that remain non-binding'. This chapter argues that gradualism and fragmentation are not entirely negative: Experimentation has paved the way for new policies, contributing to a quality check of policies with a smaller impact area, which have then been scaled up. In the wind industry, the concession rounds amounted to 43 percent of the total installed wind power capacity by the end of the last centrally given concession, meaning that they were not the largest source of turbine installations at that point. However, the concession rounds predated the Renewable Energy Law and provided useful experience for fine-tuning the legislative measures. These experiences from the wind industry were surely also beneficial to the solar PV industry, as the FiT was established there as well in 2011. What is more, the fragmented nature of Chinese governance in the solar PV industry is what led local governments to develop their solar PV manufacturing industries with exceptional speed until 2012, and the domestic PV installations between 2015 and until 31 May, 2018, when the government again centralised and put a cap on (subsidy-seeking) solar PV project approvals (Bin 2018). In particular, as Chapter 7 explains further, the growth of distributed PV installations grew quickly.

To summarise, as shown in Table 4.1, there are both drawbacks to and benefits from fragmentation and centralisation. The successful orchestration of both these characteristics of China's policy governance is what determines the final outcome.

Table 4.1 Impacts of fragmentation and centralisation in the wind and solar industries

	Benefit	*Disadvantage*
Fragmentation	Rapid growth	Overcapacity, low grid connection
Centralisation	Higher legitimacy of industry	Lower quality

Drawing together the insights on experimentation and the 'portfolio approach', as introduced in the previous chapter, we see that the Chinese institutional framework is able to draw from established actors and networks, as well as to create avenues for new ventures. These avenues often take the shape of experimentation with policies, which are later scaled up. Fragmentation pervades China's governing institutions, yet at the very apex, the CCP, by means of steering and guiding, shapes the direction and pace of new industry developments. Some impulses emanate from fragmentation and help induce change, and some originate from centralisation, drawing on hierarchy and accumulated status. The legitimacy held by state-owned enterprises (SOEs), the invaluable experience accumulated in established firms, and the alluring prospects of political careers are all contingent on preexisting authority. These processes are more inert and are capable of constraining as much as aiding change.

Fragmentation in China's energy governance has allowed for a fast-growing wind turbine and solar PV market. In times when industry development was sorely needed in order to create domestic wind turbine manufacturers, local governments were allowed to approve wind farm projects, and the media and other actors focused solely on noncritical issues with development. In times of overcapacity, the tune changed radically. This was especially evident after 2011, when the full force of centralising power was levied onto the industry and the wind industry growth rate declined, and after 2018 for the solar PV industry. The institutional traits inducing the rapid growth of China's wind power industry are based on legitimacy, alignment of expectations, and visions of incumbent and upcoming actors. I return to the role of visions and expectations in China's offshore wind industry in Chapter 6. The bottom line is that China has managed to leverage space for wind and solar energy, and the processes behind the change from fragmentation to alignment have been dominated by considerable policy flexibility. In the wind industry, the government induced policy experiments, which set in motion some of the large SOEs. In the solar PV industry, the central government remained passive until it was really needed that they stepped in, around 2011. Before that, local governments around China were quite supportive of the PV industry, indicating the importance of the fragmented nature of China's policy governance. Based on these two different industry experiences we can conclude that the CCP takes strategic advantage of fragmentation and coordination to steer the pace of development in the wind turbine and solar PV industries.

These conclusions have implications for other industries as well. The offshore wind industry in China experienced a similar stalemate between 2010 and 2017 to that faced by the onshore industry in 2006, which I will describe in Chapter 6. Furthermore, China's quest for rapid growth has come at the expense of a concern for high-quality products. For instance, the development goals set by the government have consistently been measured in terms of installed capacity, and not in terms of total electricity generated and delivered to the grid. A lack of incentive to ensure long-term electricity generation permeates the industry, impacting the chain from component suppliers to local governments approving wind and solar PV farms, SOEs investing in these farms and grid-utilities managing them. The next chapter looks further into how these institutional traits influence renewable energy industry development in China, interactions with international actors, knowledge and understanding of quality and risk, and the extent to which there is a trade-off between rapid catch-up and sufficient quality control.

Notes

1 In this book, I use *yuan* and RMB ('renminbi') interchangeably to refer to the Chinese currency.

2 The way in which interrelations between companies within the same industry conglomerate SOE-group impact quality is explained further in Chapter 6 and in Korsnes (2016).

References

Andrews-Speed, P., 2011. The institutions of energy governance in China. In William C. Ramsay and Jacques Lesourne (eds) *Chinese Climate Policy: Institutions and Intent*. Paris: Institut Français des Relations Internationales (IFRI), pp. 9–65.

Andrews-Speed, P. and Zhang, S., 2019. *China as a Global Clean Energy Champion. Lifting the Veil*. Singapore: Palgrave Macmillan.

Bin, L. 2018. China's solar industry is at a crossroads. Available at: www.chinadialogue.net/article/show/single/en/10775-China-s-solar-industry-is-at-a-crossroads (Accessed 1 May 2019).

Binz, C. and Anadon, L.D., 2018. Unrelated diversification in latecomer contexts: emergence of the Chinese solar photovoltaics industry. *Environmental Innovation and Societal Transitions*, 28, pp. 14–34.

Binz, C., Gosens, J., Hansen, T. and Hansen, U.E., 2017. Toward technology-sensitive catching-up policies: insights from renewable energy in China. *World Development*, 96, pp. 418–437.

BNEF 2010. Joined at the hip: the US-China clean energy relationship, White Paper, Bloomberg New Energy Finance (BNEF), Available at: www.bnef.com/WhitePapers/download/27 (Accessed 7 August 2011).

Chen, X. and Li, I., 2019. Beijing halves electricity bill surcharge to zap flagging economy. Available at: www.caixinglobal.com/2019-04-08/beijing-halves-electricity-bill-surcharge-to-zap-flagging-economy-101401779.html (Accessed 1 May 2019).

Cherni, J.A. and Kentish, J., 2007. Renewable energy policy and electricity market reforms in China. *Energy Policy*, 35(7), pp. 3616–3629.

china.org.cn, 2007. Medium and long-term development plan for renewable energy in China (Abbreviated Version). September, 2007. National Development and

Reform Commission (NDRC). Available at: www.china.org.cn/e-news/news070904-11.htm (Accessed 26 June 2019).

chinaenergyportal.org, 2019a. 2018 PV installations utility and distributed by province. Available at: https://chinaenergyportal.org/en/2018-pv-installations-utility-and-distributed-by-province/ (Accessed 2 May 2019).

chinaenergyportal.org, 2019b. 2018 wind power installations and production by province. Available at: https://chinaenergyportal.org/2018-wind-power-installations-and-production-by-province/ (Accessed 2 May 2019).

CNRED, 2011. *Study Report: Existing Foundations for the Establishment of a National Renewable Energy Centre*, Sino-Danish Renewable Energy Development Programme (CNRED). Retrieved at www.cnred.org.cn/english/manageInfo.do?action=showColumn&type=article&typeId=19 (Accessed 17 May 2011).

CREO, 2018. China renewable energy outlook 2018. Report prepared by the Energy Research Institute of Academy of Macroeconomic Research/NDRC and the China National Renewable Energy Centre.

CWPO, 2010. China wind power outlook 2010. Global Wind Energy Council. Available at: www.gwec.net/fileadmin/documents/test2/wind%20report0919.pdf (Accessed 05 September 2011).

CWPP, 2010. *Wind Power in China 2008: An Analysis of the Status Quo and Perspectives for Development* (2nd ed.). China Wind Power Project and Gesellschaft für Technische Zusammenarbeit (GTZ). Available at: www.cwpc.cn/cwpc/en/system/files/documents/CWPP%20Study_Wind%20Power%20in%20China%202008_Feb2010_ed.pdf (Accessed 27 September 2011).

Dai, Y. and Shi Z., 1999. Technological innovation of the Chinese photovoltaic industry. National Research Center for S&T for Development, Prepared for the Center for the Integrated Study of the Human Dimensions of Global Change, Carnegie Mellon University.

de la Tour, A., Glachant, M., and Ménière, Y., 2011. Innovation and international technology transfer: the case of the Chinese photovoltaic industry. *Energy Policy*, 39(2), pp. 761–770.

DEC, 2012. *About DEC*. Available at: www.dongfang.com.cn/index.php/aboutdecs/ (Accessed 13 March 2012).

DNV, 2011. Interview with GDUPC's Mr Xiao Jinsong: Chinese wind power industry expanding globally. *Wind Energy Update*, 2. Available at: www.dnv.com/industry/energy/publications/updates/wind_energy/2011/Windenergy_3_2011/InterviewwithGDUPCsMrXiaoJinsongChinesewindpowerindustryexpandingglobally.asp (Accessed 27 March 2012).

FS-UNEP, 2012. *Global Trends in Renewable Energy Investment 2012*. United Nations Environment Programme (UNEP), Frankfurt School of Finance and Management (FS) and Bloomberg New Energy Finance (BNEF). Available at: http://fs-unep-centre.org/publications/global-trends-renewable-energy-investment-2012 (Accessed 12 April 2012).

Gallagher, K.S., 2014. *The Globalization of Clean Energy Technology: Lessons from China*. Cambridge: MIT Press. Kindle edition.

Gan, L., 1998. *Wind Energy Development and Dissemination in China: Prospects and Constraints in an Institutional Context*. Center for International Climate and Environmental Research—Oslo (CICERO), Working Paper 1998:7. Retrieved at www.cicero.uio.no/media/178.pdf (Accessed 14 March 2012).

García, C., 2011. Grid-connected renewable energy in China: policies and institutions under gradualism, developmentalism, and socialism. *Energy Policy*, 39(12), pp. 8046–8050.

García, C., 2013. Policies and Institutions for grid-connected renewable energy: "best practice" and the case of China. *Governance*, 26(1), pp. 119–146.

Gosens, J. and Lu, Y., 2013. From lagging to leading? Technological innovation systems in emerging economies and the case of Chinese wind power. *Energy Policy*, 60, pp. 234–250.

Gosens, J., Kåberger, T. and Wang, Y., 2017. China's next renewable energy revolution: goals and mechanisms in the 13th Five Year Plan for energy. *Energy Science & Engineering*, 5(3), pp. 141–155.

Hayashi, D., Huenteler, J. and Lewis, J.I., 2018. Gone with the wind: A learning curve analysis of China's wind power industry. *Energy Policy*, 120, pp. 38–51.

Howell, T., Noellert, W.A., Hume, G., and Wolff, A.W., 2010. China's promotion of the renewable electric power equipment industry: hydro, wind, solar, biomass. Dewey & LeBoeuf LLP for the National Foreign Trade Council. Available at: www.nftc.org/default/Press%20Release/2010/China%20Renewable%20Energy.pdf (Accessed 19 May 2011).

Huang, C., Su, J., Zhao, X., Sui, J., Ru, P., Zhang, H. and Wang, X., 2012. Government funded renewable energy innovation in China. *Energy Policy*, 51, pp. 121–127.

Huang, P., Negro, S.O., Hekkert, M.P. and Bi, K., 2016. How China became a leader in solar PV: An innovation system analysis. *Renewable and Sustainable Energy Reviews*, 64, pp. 777–789.

Huo, M.L. and Zhang, D.W., 2012. Lessons from photovoltaic policies in China for future development. *Energy Policy*, 51, pp. 38–45.

Jiang, L. (ed.)., 2011. *Integrated Solution Strategies for Coordinated Wind Power and Grid Development—International Experience and China Practises.* Beijing: State Grid Energy Research Institute and Vestas Wind Technology (China) Co. Ltd.

Jiang, L., Chi, Y., Qin, H., Pei, Z., Li, Q., Liu, M., Bai, J., Wang, W., Feng, S., Kong, W. and Wang, Q., 2011. Wind energy in China. Status and Prospects. *IEEE Power and Energy Magazine*, 9(6), pp. 36–46.

Korsnes, M., 2016. A sustainable Chinese catch-up? Product quality and interactive learning in the offshore wind industry. *International Journal of Technological Learning, Innovation and Development*, 8(2), pp. 172–196.

Lema, A., and Ruby, K., 2007. Between fragmented authoritarianism and policy coordination: Creating a Chinese market for wind energy. *Energy Policy*, 35(7), pp. 3879–3890.

Lew, D.J., 2000. Alternatives to coal and candles: wind power in China. *Energy Policy*, 28(4), pp. 271–286.

Lewis, J., 2007. Technology acquisition and innovation in the developing world: wind turbine development in China and India. *Studies in Comparative International Development*, 42(3–4), pp. 208–232.

Lewis, L., 2011. China's competitive advantage in the solar industry: how advantageous is it really?, Chinafaqs.org, Available at: www.chinafaqs.org/blog-posts/chinas-competitive-advantage-solar-industry-how-advantageous-it-really (Accessed 5 October 2011).

Lewis, J., 2013. *Green Innovation in China—China's Wind Power Industry and the Global Transition to a Low-Carbon Economy*, New York: Columbia University Press.

Lewis, J. and Wiser, R.H., 2007. Fostering a renewable energy technology industry: An international comparison of wind industry policy support mechanisms. *Energy Policy*, 35(3), pp. 1844–1857.

Li, H., Meng, L., Wang, Q., and Zhou, L.A., 2008. Political connections, financing and firm performance: evidence from Chinese private firms, *Journal of Development Economics*, 87(2), pp. 283–299.

Li, J., Shi, J., Xie, H., Song, Y. and Shi, P., 2006. *A Study on the Pricing Policy of Wind Power in China*, Chinese Renewable Energy Industry Association, Greenpeace, Global Wind Energy Council. Available at: http://gwec.net/wp-content/uploads/2012/06/Report-wind-power-price-policy-china.pdf (Accessed 11 September 2011).

Li, J., Gao, H., Shi, P., Shi, J., Ma, L., Qin, H., and Song, Y., 2007. *China Wind Power Report 2007*. Beijing: China Environmental Science Press. Available at: http://gwec.net/wp-content/uploads/2012/08/wind-power-report.pdf (Accessed 11 September 2011).

Li, J., Cai, F., Qiao, L., Xie, H., Gao, H., Yang, X., Tang, W., Wang, W., Li, X., 2012. China wind energy outlook 2012. Chinese Renewable Energy Industry Association, Greenpeace, Global Wind Energy Council, Chinese Wind Energy Association, Beijing.

Lin, B., 2018. China is a renewable energy champion. But it's time for a new approach. Available at: www.weforum.org/agenda/2018/05/china-is-a-renewable-energy-champion-but-its-time-for-a-new-approach/ (Accessed 1 May 2019).

Liu, J., 2019. China's renewable energy law and policy: a critical review. *Renewable and Sustainable Energy Reviews*, 99, pp. 212–219.

Mathews, J.A. and Tan, H., 2015. China's green-energy revolution. *project syndicate*. Available at: www.project-syndicate.org/commentary/china-green-energy-revolution-by-john-a-mathews-and-hao-tan-2015-05 (Accessed 16 May 2015).

Marigo, N., 2007. The Chinese silicon photovoltaic industry and market: a critical review of trends and outlook. *Progress in Photovoltaics: Research and Applications*, 15, pp. 143–162.

Marigo, N., Foxon, T.J., and Pearson, P.J.G., 2010. Chinese low-carbon innovation: developing technological capabilities in the solar photovoltaic manufacturing industry. *Journal of Knowledge-based Innovation in China*, 2(3), pp. 253–268.

Martinot, E., 2010. Renewable power for China: past, present, and future. *Frontiers of Energy and Power Engineering in China*, 4(3), pp. 287–294.

Martinot, E. and Li, J., 2007. *Worldwatch Report. Powering China's Development: the Role of Renewable Energy*. Washington, DC: The Worldwatch Institute.

Martinot, E. and Li, J., 2010. Renewable energy policy update for China. *Renewable Energy World*, 21 July. Available at: www.renewableenergyworld.com/rea/news/article/2010/07/renewable-energy-policy-update-for-china (Accessed 5 March 2011).

Mingyang, 2011. *2010 Annual Report*. China Ming Yang Wind Power Group Limited. Available at: http://media.corporate-ir.net/media_files/IROL/23/238508/2010_Annual_Report.pdf (Accessed 11 January 2012).

Nahm, J. and Steinfeld, E.S., 2014. Scale-up nation: China's specialization in innovative manufacturing. *World Development*, 54, pp. 288–300.

Nahm, J., 2017. Exploiting the implementation gap: policy divergence and industrial upgrading in China's wind and solar sectors. *The China Quarterly*, 231, pp. 705–727.

Naughton, B., 2010. Chapter 4. Economic growth: from high-speed to high-quality. In J. Fewsmith (ed.) *China Today, China Tomorrow: Domestic Politics, Economy, and Society*. Lanham, MD: Rowman & Littlefield.

Naughton, B., 2017. Is China socialist? *Journal of Economic Perspectives*, 31(1), pp. 3–24.

Publicover, B., 2019. PV plants in China under greater pressure than wind farms. April 23, 2019. Available at: www.pv-magazine.com/2019/04/23/pv-plants-in-china-under-greater-pressure-than-wind-farms-fitch/ (Accessed 29 April 2019).

Quitzow, R., 2015. Dynamics of a policy-driven market: The co-evolution of technological innovation systems for solar photovoltaics in China and Germany. *Environmental Innovation and Societal Transitions*, 17, pp. 126–148.

Quitzow, R., Huenteler, J. and Asmussen, H., 2017. Development trajectories in China's wind and solar energy industries: How technology-related differences shape the dynamics of industry localization and catching up. *Journal of Cleaner Production*, 158, pp. 122–133.

RED, 2011. Sino-Danish renewable energy development programme, 'The renewable energy industrial development report 2010'. Available at: www.cnred.org.cn/english/webSite/downLoad.jsp?fileName=RE+center+Pre-FS-task+1-English_20110412011426.pdf (Accessed 17 May 2011).

REN21, 2009. *Background Paper: Chinese Renewables Status Report*. Renewable Energy Policy Network for the 21st Century (REN21) and CREIA. Available at: www.ren21.net/REN21ProductsServices/Publications/RegionalStatusReports/RenewableEnergyinChina/tabid/5441/Default.aspx (Accessed 19 May 2011).

REN21, 2011. *Renewables 2011 Global Status Report*. Renewable Energy Policy Network for the 21st Century (REN21). Available at: www.ren21.net/Portals/97/documents/GSR/GSR2011_Master18.pdf (Accessed 10 August 2011).

reuters.com, 2011. China to slow solar output boom as price slips, by Leonora Walet. Available at: www.reuters.com/article/2011/11/02/us-china-solar-siew-idUSTRE7A14QD20111102 (Accessed 2 November 2011).

reuters.com, 2013. Creditors of Suntech's main unit back debt restructuring plan. Available at: www.reuters.com/article/us-suntech-shunfeng-creditors/creditors-of-suntechs-main-unit-back-debt-restructuring-plan-idUSBRE9AB0GD20131112 (Accessed 31 May 2019).

Rosen, D., and Houser, T., 2007. *China Energy: A Guide for the Perplexed*. Center for Strategic and International Studies and the Peterson Institute for International Economics, online: www.iie.com/publications/papers/rosen0507.pdf (Accessed 16 May 2011).

Ru, P., Zhi, Q., Zhang, F., Zhong, X., Li, J. and Su, J., 2012. Behind the development of technology: The transition of innovation modes in China's wind turbine manufacturing industry. *Energy Policy*, 43, pp. 58–69.

Saich, T., 2011. *Governance and Politics of China* (3rd ed.). Houndmills, Basingstoke, Hampshire/New York: Palgrave Macmillan.

Schuman, S. and Lin, A., 2012. China's renewable energy law and its impact on renewable power in China: Progress, challenges and recommendations for improving implementation. *Energy Policy*, 51, pp. 89–109.

Shubbak, M.H., 2019. The technological system of production and innovation: The case of photovoltaic technology in China. *Research Policy*, 48(4), pp. 993–1015.

Urban, F., Nordensvärd, J. and Zhou, Y., 2012. Key actors and their motives for wind energy innovation in China. *Innovation and Development*, 2(1), pp. 111–130.

Walter, C.E., and Howie, F.J.T., 2011. *Red Capitalism: The Fragile Financial Foundation of China's Extraordinary Rise*. Singapore: Wiley-Blackwell.

Wang, M., 2007. Issues related to the implementation of China's energy law: analysis of the energy conservation law and the renewable energy law as examples. *Vermont Journal of Environmental Law*, 8, pp. 225–250.

Wang, Q., 2010. Effective policies for renewable energy—the example of China's wind power—lessons for China's photovoltaic power. *Renewable and Sustainable Energy Reviews*, 14(2), 702–712.

Wang, Z., Qin, H. and Lewis, J.I., 2012. China's wind power industry: policy support, technological achievements, and emerging challenges. *Energy Policy*, 51, pp. 80–88.

WWEA, 2014. Statistics of wind power development in China 2013. Available at: *Quarterly Bulletin*, 1, World Wind Energy Association (WWEA), online: www.wwindea.org/webimages/WWEA_Bulletin-ISSUE_1_2014_reduced.pdf (Accessed 30 June 2014).

XEMC-Darwind, 2012. About us. Available at: www.xemc-darwind.com/index.php/about-us.html (Accessed 11 March 2012).

Xu, C., 2011. The fundamental institutions of China's reforms and development. *Journal of Economic Literature*, 49(4), pp. 1076–1151.

Ydersbond, I.M. and Korsnes, M., 2016. What drives investment in wind energy? A comparative study of China and the European Union. *Energy, Research & Social Science*, 12, pp. 50–61.

Ye, L.C., Rodrigues, J.F. and Lin, H.X., 2017. Analysis of feed-in tariff policies for solar photovoltaic in China 2011–2016. *Applied Energy*, 203, pp. 496–505.

Yi-chong, X. 2017. *Sinews of Power. The Politics of the State Grid Corporation of China*. New York: Oxford University Press.

Yuan, X., and Zuo, J., 2011. Transition to low carbon energy policies in China—from the Five-Year Plan perspective. *Energy Policy*, 39(6), pp. 3855–3859.

Yue, L., 2012. Dynamics of clean coal-fired power generation development in China. *Energy Policy*, 51, pp. 138–142.

Zhang, F. and Gallagher, K.S., 2016. Innovation and technology transfer through global value chains: Evidence from China's PV industry. *Energy Policy*, 94, pp. 191–203.

Zhang, S., Andrews-Speed, P. and Zhao, X., 2013. Political and institutional analysis of the successes and failures of China's wind power policy. *Energy Policy*, 56, pp. 1–10.

Zhang, S., Andrews-Speed, P., Zhao, X. and He, Y., 2013. Interactions between renewable energy policy and renewable energy industrial policy: A critical analysis of China's policy approach to renewable energies. *Energy Policy*, 62, pp. 342–353.

Zhang, W. and White, S., 2016. Overcoming the liability of newness: Entrepreneurial action and the emergence of China's private solar photovoltaic firms. *Research Policy*, 45(3), pp. 604–617.

Zhao, Z.-Y., Hu, J., and Zuo, J., 2009. Performance of wind power industry development in China: A diamond model study. *Renewable Energy*, 34(12), pp. 2883–2891.

Zhou, Y., Zhang, B., Zou, J., Bi, J. and Wang, K., 2012. Joint R&D in low-carbon technology development in China: A case study of the wind-turbine manufacturing industry. *Energy Policy*, 46, pp. 100–108.

Zhou, Y., Pan, M. and Urban, F., 2018. Comparing the international knowledge flow of China's wind and solar photovoltaic (pv) industries: Patent analysis and implications for sustainable development. *Sustainability*, 10(6), 1883.

Zhu, M., Qi, Y., Belis, D., Lu, J. and Kerremans, B., 2019. The China wind paradox: The role of state-owned enterprises in wind power investment versus wind curtailment. *Energy Policy*, 127, pp. 200–212.

5 The international relations of the wind and solar industries

Markets, experiences and initiatives outside of China have impacted China's solar PV and wind industries in considerable but different ways. As established in the previous chapter, China's wind industry was mainly based on the domestic market, and licensed technologies from European companies, whilst the solar PV industry to a large extent was supplying PV panels to the European and US markets and produced the panels domestically with turnkey industry plants from European and US companies. Only around 2012/2013 did China for real start installing PV domestically, around the same time as Germany started reducing government support for PV. This chapter shows the interplay, interrelations and negotiations occurring between China and the world during the buildup of the wind and solar industries. Particular attention is paid to foreign actors on the one hand who want to sell their knowledge and gain access to the Chinese markets, and on the other hand, Chinese actors who want to learn but also prefer to do things their own way.

In 2010, China captured about 50 percent of the global wind turbine market, up from 4.4 percent in 2005, and almost all of these turbines were installed domestically (REN21 2011). In the solar PV market, the statistics were surprisingly similar except that most PV panels were exported: China (excl. Taiwan) produced about 50 percent of the global solar cells in 2010, compared to about 5 percent in 2004 (Marigo 2007). From 2010, this changed abruptly, and by the end of 2017, China had already installed 33 percent of the world cumulative PV domestically (131 gigawatts [GW] out of a total of 402 GW), at the same time as it manufactured about 70 percent of global solar cell and module supply (Bin 2018; REN21 2018). For wind energy, China continued to lead the world in wind turbine manufacturing by the end of 2017 and captured about 35 percent of the accumulated global wind power capacity, but most turbines were still installed domestically (CWEA 2018). Some Chinese wind turbine companies such as Goldwind, Mingyang and Envision have ventured into international markets, cultivating interactions with countries such as India, Thailand and Ethiopia, but this is still just a marginal aspect of their business (Zhou, Pan and Urban

2018). Let's take a closer look at the role of international experiences for China's wind and solar PV industries.

International experiences of the wind industry

As has been excellently documented in Lewis' (2013) book *Green Innovation in China*, and the more recent Kirkegaard (2019) book *Wind Power in China*, the Chinese wind power industry was from the outset highly dependent on, and benefitted from the experiences of, international wind turbine manufacturing companies such as NEG Micon (which later merged with Vestas), Vestas, Gamesa, Nordex and GE Wind. The relationships started out with imports of whole turbines and went through different types of partnerships, to local Chinese original equipment manufacturing. Important learning networks for the Chinese wind industry were established with foreign companies through licenses or joint development (Lewis 2007; 2011). Regional and global learning networks have been important in the wind industry since the outset, and emerging markets in China and other Asian countries have served as 'valuable regional learning networks for new firms' (Lewis 2011, p. 283). These new markets, in turn, have made an important contribution to 'global learning networks of knowledge and innovation', because of the global character of the wind industry (Lewis 2011, p. 283). The interactions between China and the international wind industry can therefore characterised as interdependent, and not just a one-way knowledge transfer to China (Lewis 2013).

Several of the Chinese wind turbine manufacturers made licensing agreements with German, Danish and US firms. The nature of these networks varied, and what type of information that was shared depended on the agreement. Lewis (2011, p. 296) maintains that these networks can be both positive and negative, because on one hand they can 'increase access to global learning and experience worldwide, which is likely beneficial', whilst on the other hand they can 'create competitors and make it harder to safeguard valuable or sensitive information'. According to Nahm (2017, p. 713), of the 31 largest wind turbine manufacturers, '16 entered license agreements with foreign firms, 14 signed joint-development contracts, six autonomously developed wind turbine technologies, and three started joint venture operations'. When Goldwind acquired a majority stake in the German company Vensys in 2008, it became engaged with research and development activities in Germany, although the company until then had been uniquely focussed on China (Lewis 2011, p. 297). Several Chinese wind turbine manufacturers also established research centres in leading wind turbine research and development communities globally, such as at Risøe in Denmark (e.g., Mingyang and Envision). This enabled these companies to tap into highly valuable knowledge or skilled personnel, crucial to further develop their turbine designs.

Another important learning network for Chinese manufacturing firms was more regional and was found at the sites where both Chinese and international firms performed their wind turbine testing in China. Lewis (2007) illustrates this with the case of Goldwind, which has placed its testing operations in a popular wind farm site where almost all of the leading global wind turbine manufacturers in China have also installed turbines. Goldwind benefitted from the learning shared amongst 'these manufacturers coming from the other wind learning hubs of the world, as they all tested their designs in China' (Lewis 2007, p. 228). The same can be assumed about most large wind turbine manufacturers in China. What is more, several foreign wind turbine suppliers also established local production facilities, and foreign wind turbine manufacturers also sourced components from Chinese suppliers (Nahm 2017). Compared to European firms, which are more vertically integrated, component supply for wind turbines in China has a shared supply base of components, allowing for industrialisation, quick manufacturing and economies of scale (Lema, Berger and Schmitz 2013). In 2013, there were 'more than 50 blade manufacturers, 100 tower manufacturers and 1,000 manufacturers producing other components and parts for the onshore wind industry', and these were both domestic and international companies (Korsnes 2016, p. 182). Therefore, foreign wind turbine component suppliers gained market shares in China, compared to foreign turbine manufacturers which lost market shares (Lema, Berger and Schmitz 2013). This implies that international networks were effective both locally and globally, meaning that they shared knowledge, in terms of licensing agreements and other encounters, and they shared experiences and routines, from which Chinese manufacturers would learn. Foreign knowledge has been central in building up the Chinese industry, implying that shortcomings in China's domestic innovation capabilities can be compensated by international connections (Binz *et al.* 2012). In this sense, as we will soon see, China's wind and solar PV industries share similarities in terms of development.

There have also been several wind energy projects and bilateral cooperation initiatives on a governmental level, as well as between universities and research institutes. Some examples of cooperation between China and the Nordic countries are provided by Delman and Chen (2008): the Danish wind energy development programme (WED) in cooperation with the National Development and Reform Commission (NDRC), the Danish-Chinese Renewable Energy Programme (RED) also located at the NDRC, the Swedish-Chinese exchange programme on renewable energy between the Royal Swedish Academy of Engineering Sciences (IVA) and the Chinese Academy of Engineering (CAE) and the EU-China Energy and Environment Programme (EEP). In addition, several other countries have had activities within the renewable energy field in China, such as France, Germany, Japan, the Netherlands, Spain, the United Kingdom and the United States. For example, Germany's GIZ, a federal enterprise which supports the German government in international cooperation for sustainable development, had a wind power project in China between 2005 and 2010, which strived to 'improve the technical capabilities of private and state institutions for the nationwide development of grid-bound wind energy use' in China (CWPP

2010, p. 12). One of their core tasks was to educate personnel and managers, since China was lacking this particular expertise within the wind sector (CWPP 2010).

Already around 2011, Chinese wind turbine manufacturers were increasingly seeking to export their turbines. Overcapacity and increased competition domestically induced this trend. This was, for example, voiced by Mr Xiao Jinsong, deputy director of Guodian United Power's Chief Engineering Office: 'As an old Chinese saying goes, distant water won't help to put out a fire close at hand. So, exploiting the overseas market has become the only way out for Chinese manufacturers' (DNV 2011, p. 30). Nevertheless, as we see now in 2019, Chinese turbine manufacturers are still not exporting large quantities. There has been a focus on cultivating relationships with countries which are less competitive but lucrative, such as India, Cuba, Romania, Panama, Thailand and Ethiopia, but Chinese firms still lack shares in international markets (Koch-Weser and Meick 2015; Zhou, Pan and Urban 2018). This could be related to a failure in ensuring that Chinese turbines are of high quality, that is, in proving the legitimacy of Chinese wind turbines internationally. Many of the larger Chinese manufacturers have been interested in expanding internationally, but buyers have been reluctant due to 'thinner documentation' of their wind turbine performance, as well as questions related to operation service and maintenance abroad (Rechargenews.com 2011). Moreover, in the United States, being a Chinese state-owned company 'carries heavy political baggage' and could have serious implications for the legitimacy of Chinese actors there (BNEF 2010, p. 13). Some of the projects which have been installed abroad were serving as demonstration projects. For instance, Sinovel erected two test turbines in Sweden with the stated purpose of a 'demonstration effect': 'If Sinovel turbines are accepted by local users for the quality and technology, and meet no problem in access to the grid, it will pave the way for its turbines to enter the European market' (Sinovel.com 2012). Since then, however, Sinovel has faced severe difficulties and has been inactive for several years due to a combination of overcapacity, bad management and a litigation case with their US provider of control systems, accused of stealing trade secrets (Kirkegaard 2019).

International experiences of the solar PV industry

The solar PV industry has become a veritable global industry. This, for instance, is evident when looking at the role of the German market in initiating the Chinese solar PV industry, and the increase in Chinese domestic installations of PV panels simultaneously with the cut-back of the German solar PV FiT. The Chinese solar PV industry started out through the aid of several conjunctures which were of international nature (Green 2019). China joined the World Trade Organization (WTO) in 2001, and this greatly benefitted the Chinese PV first movers. For instance, as stated by the initiator of LDK Solar (founded 2005): 'if China had not become a member of the WTO, there would have been no market for our PV products and I might not have founded LDK Solar' (cited in Huang *et al.* 2016, p. 781). In 2002, private entrepreneurs were officially allowed to become members of the CCP, which elevated the status of private enterprises—making it more interesting for Chinese both abroad and domestically to start their own companies in China (Dickson 2007; Green 2019).[1] What is more, several of the CEOs and staff of the solar PV startups were educated abroad, or actually started the company abroad and later moved main parts to China (as the case with Canadian Solar was). We will now take a closer look at the various ways in which international experiences, actors and events impacted the Chinese solar PV industry.

Binz and Anadon (2018, p. 28) argue that the most important comparative advantage for Chinese solar PV actors was 'their international connections, the quick upscaling of operations with support from local governments and the fast emergence of a low-cost local supplier industry'. International trade in solar technology production equipment was one of the central ways which the Chinese solar PV industry acquired technology and competences, making process-related aspects of the technological learning process important (Quitzow, Huenteler and Asmussen 2017). The international experiences of the group of CEOs and staff from the first and second waves of entrants to the Chinese solar PV industry between 1997 and 2007 have been recognised as highly important for the initial success of these companies (i.e., de la Tour *et al.* 2011; Zhang and Gallagher 2016). The importance of this labour mobility was already described in the previous chapter. Nevertheless, there were several other interrelations between the global PV industry and China which led to the emergence of the PV industry. Incidents outside of China were triggering factors for the emergence a Chinese domestic manufacturing industry as well as the emergence of domestic PV installations.

One important factor was access to finance from international investors, since domestic Chinese support was scarce and insufficient once the PV companies started growing more quickly around 2004. Suntech was approached by American investors already around 2004 and proposed the idea of an initial public offering (IPO) or a stock market launch, which would give the company more money (Binz and Anadon 2018). In order for this to happen, they first needed to buy out the state-owned investors (Binz and Anadon 2018). Then, in December 2005, Suntech became the first private Chinese-based company to be listed on the New York Stock Exchange, indicating that the Chinese PV manufacturers were promising and interesting to international investors (Huang *et al.* 2016; Green 2019). Within one and a half years, seven other Chinese solar PV firms were listed in international stock markets (Zhang and White 2016). Even though some of these companies, such as Trina and Canadian Solar, were doing basic module assembly and not solar cell manufacturing, they benefitted from Suntechs success (Green 2019). The additional money received from these international investments were very useful to manifest the promise of the industry, and shortly after Chinese public and private investors were also investing in the domestic PV companies (Binz and Anadon 2018). This access to global capital markets was 'a key trigger for the subsequent fast expansion of the Chinese PV industry' (Binz and Anadon 2018, p. 27).

During the 2008/2009 financial crisis, the global PV industry was hit hard. However, at that point, the Chinese PV industry had become too important for the Chinese government to lose. The industry was declared a 'strategic emerging industry' (along with the wind industry as well), and Chinese solar PV firms (and wind firms) therefore benefitted from the 4,000-billion-*yuan* Chinese stimulus package, and domestic development banks were supporting the industry (Quitzow 2015; Huang *et al.* 2016). The stimulus package resulted in the Golden Sun Demonstration Programme, and the Solar Building Programme launched in 2009 (Huang *et al.* 2016). Soon after, a feed-in tariff came in place, as was described in the previous chapter. The timing of entry from the Chinese central government appears to have been excellent. At the time of entry, the prices of PV panels had already fallen considerably, the Chinese PV manufacturing industry was facing severe problems with overcapacity, international markets were in decline, and complaints of dumping were emerging from the European Union and the United States. It is difficult to think that this time of entry of the Chinese government was only based on luck; surely, there was never really a great need for the government to step in before that point in time.

Coinciding with the huge Chinese manufacturing expansion was also a global industry which profited from delivering equipment and goods for manufacturing to China. For example, in 2010, China exported around 15 to 20 percent of solar PV products to the United States (Chinadaily.com.cn 2011), but the United States was still a net exporter of solar technology to China. Their largest export was polysilicon (used to make crystalline silicon PV panels), and the second-largest solar energy export product was manufacturing equipment for PV products (GTM report 2011). PV manufacturers around the world lost out in competition with the cheaper Chinese solar panels, and around 2011 there were several bankruptcies. In the United States alone, at least three companies went bankrupt, including Solyndra and Evergreen Solar Inc., which were companies praised for their innovative technology and who had benefitted from the US financial aid package after the financial crisis (nytimes.com 2011; Shubbak 2019). Many claimed that China was to be blamed for these bankruptcies because of lower prices and the Chinese firms' cheap access to capital (Lacey and Romm 2011). Also, allegations were made against the Chinese manufacturers regarding their selling of products below cost in order to gain market share (referred to as dumping). This is what spurred a trade complaint filed by seven American solar cell and panel manufacturers with the potential repercussions of a 100 percent tariff on Chinese panels (nytimes.com 2011). One of the spokespersons behind this trade complaint voiced a fear that 'China will develop its own sources for capital equipment and raw materials and then the whole supply chain will be gone [for the U.S.]' (Renewableenergyworld.com 2011). In other words, it was not only the fact that China sold panels cheaply which was problematic for them, but also that China would develop a full industry chain, ending imports of polysilicon and manufacturing equipment from the United States. Fast-forward seven years, China had the following market shares of central solar PV products by the end of 2017: 55 percent of world polysilicon production, 87 percent of silicon wafer production, 69 percent of solar cell production and 72 percent of solar modules (Bin 2018). In a study comparing China's wind and solar innovation capabilities, Nahm and Steinfeld (2014, p. 298) come to the conclusion that the interrelatedness of the solar PV industry makes trade-policy more tricky: 'the ostensibly "made in China" technology products are the result of highly collaborative commercial innovation efforts, ones which frequently involve—and redound to the profit of—firms from the countries ostensibly being most harmed'.

With license to build: the role of certification agencies in the offshore wind industry[2]

The role of European certification firms in China's offshore wind industry development is a particularly interesting case, as it illustrates well how European actors not only were trying to sell a service, but also was preparing the grounds for the legitimacy of this service. Certification is understood as a procedure which relates to the work of both advisory and certification agencies. Both types of agencies work with international standards, but there is a small difference between the two types of agencies in way they operate. Advisory firms are able to perform and contribute with tacit and explicit knowledge and experience. Certification agencies, contrarily, merely point to issues which diverge from standards without providing advice on how to best solve an issue, thereby providing a role for advisory agencies. Certification has exploded over the past 25 years (Bartley 2011). As risk and uncertainty have become fields to be 'managed' (Power 2007), certification firms have arguably gained an important role by providing evidence that a technology will function within a given set of conditions. Moreover, as certification firms certify products and processes according to internationally acknowledged standards, they may also be seen as 'maintenance agents' of standards (Timmermans and Epstein 2010), meaning that they are not neutral agents.

By better understanding the roles and strategies of European certification firms in the development of offshore wind energy in China, this chapter explores how these strategies may impact technology development in China. Thus, China's offshore wind industry here serves as a case in point in elucidating how certification procedures are introduced and welcomed in a new setting, and how these impact innovation and technology development in China.

Certification, intermediaries and technology development

At the end of the nineteenth century, certification became important in the shipping industry as the brutal forces of nature made insurance critical for ship owners to increase safety by setting standards and implementing surveillance, and to reduce risks by making correct and detailed assessments (Andersen and Collett 1989). In order to do this, ship owners created local organisations to share technological knowledge and mutually insure their ships. These organisations codified the knowledge which skilled seafarers had of which ships were good for which purposes; subsequently, they assigned ships to different classes (hence, in the shipping industry, certification is referred to as classification) and set standards for safety and risk evaluation, making it easier for insurance premiums to be determined (Andersen and Collett 1989). Insurance is therefore closely connected with certification. At the beginning of the twentieth century, certification was rare (Busch 2011) and certificates were regarded as unnecessary, since people were generally able to assess risk based on personal experience and relationships. Perceived risks were therefore assessed in conjunction with levels of trust. Today, as choices and differentiation have increased considerably with globalisation, certificates have become commonplace in order to (seemingly) establish trust and manage risk (Busch 2011). We may therefore say that bonds of trust and assessments of risk, to a great extent, are outsourced to allegedly independent organisations which issue certificates which conform to established standards.

Risk is a central concept to certification agencies. Since risk management deals with uncertainty, risks must be rendered thinkable and attention to risks must be triggered and interpreted (Power 2007). However, risks are evaluated differently depending on criteria which are open for interpretation in different contexts (Pesendorfer 2011). Therefore, success in selling services related to risk reduction depends on the persuasive power of risk management companies in convincing others that their knowledge will make products and processes safer. Perceptions of risk and quality are closely connected, as understanding of product quality often means that there are few risks associated with the product. Nevertheless, there is no consensus on the definition of quality (Reeves and Bednar 1994; Yong and Wilkinson 2002). What is understood as 'quality' changes according to the criteria at hand and can be difficult to measure (Garvin 1984; Sousa and Voss 2002). Measuring conformity to an established standard has become an increasingly common way to measure quality (Garvin 1984; Busch 2011). This implies that quality has moved from being a concept that mainly occupied engineers to becoming a field of main interest to company managers (Power 1997). Quality, seen this way, is then a result of a process which makes 'quality' measurable, even in cases where it is difficult to measure, so it can be audited by a third party. Power (1997, p. 60) therefore finds that 'there is no quality without quality assurance', implying that quality is a form of impression management. In other words, as long as a product conforms to criteria set by someone, typically frontier firms in developed countries, it is of good quality.

But how do these frontier companies provide legitimacy for their endeavour? Legitimacy within an auditing system is nested in a circle wherein involved actors check on each other to ensure that standards are maintained (Busch 2011). In this circle, certification agencies certify according to internationally acknowledged standards, and certification agencies are, in turn, accredited (i.e., certified) by international organisations and governments (Gustafsson and Tamm Hallström 2013). As Gustafsson and Tamm Hallström (2012; 2013) have shown, certification agencies gain legitimacy by referring to accreditation agencies, and these, in turn, continue to refer to higher sources of legitimacy (governments, member organisations and international organisations). Hence, as European certification agencies attempt to access China, they must make sure that: (1) risks are assessed similarly and (2) they, as certification agencies, are acknowledged as legitimate actors to carry out the certification process to ensure product quality. This might be a rather difficult process. As pointed out by Kirkegaard (2019, p. 192):

> Instead of being based on confidence in the quality of a product through legalities, technicalities, certificate, or formalities, the quality of wind turbines and their components in China has to a large extent rather been measured in terms of 'interpersonal relations' (guānxì).

Certificates can be seen as an outsourcing of trust to external or independent third parties, which need to have substantial local legitimacy to be trusted.

The development of standards is always contingent on local work, in the sense that each location develops and interprets their own standards. Thus, standards are not automatically universal, and Timmermans and Berg (1997, p. 275) have instead referred to *local universality*, as standards emerge from 'localized processes of negotiations and pre-existing institutional, infrastructural, and material relations'. As we will see, tensions between local practices and the universality of standards are central to this chapter. A study by Ponte and Cheyns (2013) showed that complying with standards or partaking in the development of standards can be more difficult in developing countries, as such countries lack resources or are not familiar with the 'rules of the game'. Although China might not be called a developing country per se, complying with or developing standards may be more difficult for 'latecomer countries' such as China, as it has not had as much influence over and experience with the international standardisation system as Western countries (Ernst 2013). Moreover, certification agencies also take part in shaping international standards by participating in technical groups of international standardisation organisations (ISO and IEC) (Tamm Hallström 1996; 2000). Chinese certification agencies are not welcome to the 'inner circle of the good international club', and therefore also have limited opportunities to influence the outcomes of internationals standards (Kirkegaard 2019, p. 260).

In terms of technology development, certification agencies can be grouped under the more general concept of 'intermediaries' (Stewart and Hyysalo 2008). Intermediaries can be people or organisations which adapt a technology to better fit with user needs (Russell and Williams 2002) and can be said to perform three main roles: configuring, facilitating and brokering (Stewart and Hyysalo 2008). They facilitate by providing opportunities to others. They configure by providing interpretations, and by actively creating content, setting rules and defining priorities of use, and they broker by representing and raising support for spokespeople of a technology (Stewart and Hyysalo 2008). Intermediaries can play various roles in an innovation process, but generally act as brokers between two or more parties (Howells 2006) and are important in the process of facilitating learning and transferring knowledge and technology 'across people, organizations and industries' (Hargadon and Sutton 1997; Bessant and Rush 1995; Howells 2006). In China, intermediaries between Chinese and foreign actors are crucial to get things done for the entering companies. Having a local Chinese agent that greases the wheels and 'nurture[s] the soil politically and commercially' is central to get ahead in the Chinese market (Kirkegaard 2019, p. 181). In the context of certification agents, some studies have focused on the way in which certification agencies participate in the development of international standards (e.g., Higgins and Tamm Hallström 2007), and the role of consultants in innovation processes (e.g., Sturdy and Wright 2011; Wright, Sturdy and Wylie 2012). The next part of the chapter explores the way in which certification agencies as intermediaries perform their work and how they work to get clients in China, where certification services are not as common as in Europe.

The organisation of offshore wind industry certification procedures in China and Europe

'Certification' here refers to a procedure performed by an independent third party to ensure that the practices used in the production of a product conform to industry best practice; it is thus often referred to as 'conformity assessment' by the industrial world (Busch 2011). Best practice is often codified in the shape of international standards. The International Electrotechnical Commission (IEC) has, since 1987, had a technical committee (TC 88) working on setting international standards for wind turbines, the IEC 61400 series (IEC 2010). As of 2019, the committee had 30 participating countries (including China) and ten observing countries, all of which contributed to the development of standards for design requirements, engineering integrity, measurement techniques and test procedures of wind turbines (IEC 2019). These standards, in turn, are used as the basis for the certification of turbines, which can be performed by an accredited third party. Accreditation, itself, is also based on a standard (ISO/IEC 17065) issued jointly through the International Organization for Standardization (ISO) and the IEC. Certification means, therefore, that a third party has reviewed the 'suitability, adequacy and effectiveness' of a product, process or system and issued an attestation including a 'statement, based on a decision following the review, that fulfilment of specified requirements (e.g., guidelines, codes and standards) has been demonstrated' (Woebbeking 2008, p. 2).

As demonstrated by the dotted line in Figure 5.1, certification and advisory firms typically enter an industry when there is sufficient accumulated knowledge about the technology and its risks. Exactly *when* this level is reached depends on the way in which each agency assesses business opportunities (i.e., when the industry is regarded as 'auditable', as has been pointed out by Power [1996]). By assembling knowledge and standardising procedures, certification agencies write guidelines which are voluntarily adopted by industry actors. International standards, typically developed by certification agencies, governments and technical experts, often take these guidelines into account as a basis for the standard (Tamm Hallström 1996). As certification and advisory agencies enter China, where technology maturity is lower than the global norm, agencies are very aware that their experience is useful. Chinese actors, on the other hand, have less experience making it difficult to establish their own certification procedures. Their challenge is thus to avoid major risks and acquire knowledge and experience without outsourcing projects to foreign actors. Note that Figure 5.1 is meant for illustrative purposes, and does not intend to accurately represent the various elements.

Within the wind industry there is a distinction between 'type certificates' and 'project certificates'. A type certificate attests that a wind turbine and its components have been manufactured according to certain requirements. A company may also choose to get a design assessment, and this assessment is less comprehensive than a type certificate since it disregards certain stages related to testing and manufacturing the design (Woebbeking 2008). A project certificate is intended for a project with more than one turbine. It also includes certification of all of the necessary installations such as measuring masts, electricity transmitters, transformer stations and so on. A project certificate therefore ensures that the different components of a wind farm project are compatible, suitable and adequate. In Europe, project certificates are often

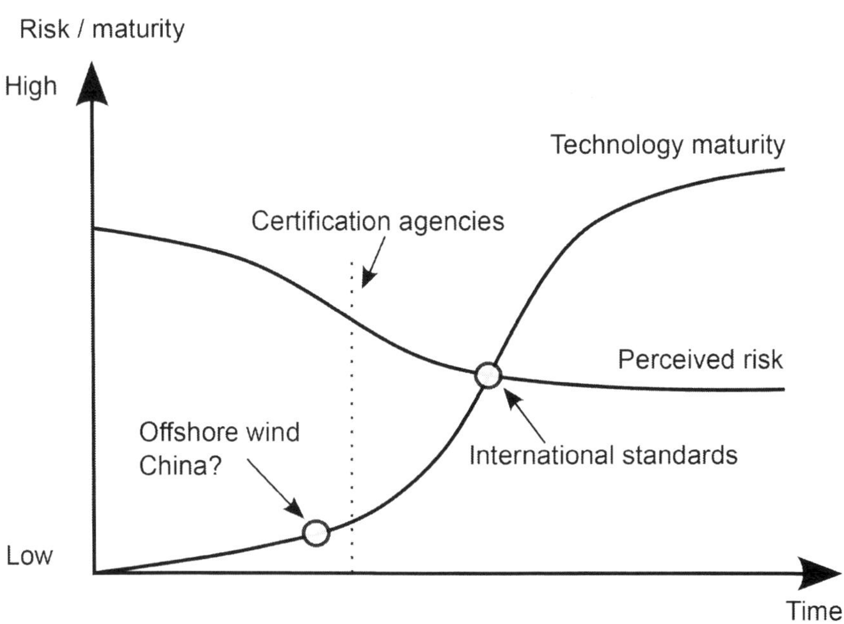

Figure 5.1 Risk, technology maturity and scope for certification agencies.

Source: Author.

necessary to get insurance, to comply with government regulations, and to attract finance. In China, however, the scope for such certificates is to a large extent covered by the government, and government-affiliated agencies.

Within China, testing designs and selling turbines is possible without an international certificate. Project developers, design institutes and construction agencies are the three most important actors. Especially design institutes such as Shanghai Investigation Design & Research Institute and HydroChina Engineering Corporation cover much of the work of European certification agencies. These agencies are government affiliated, as are most of the construction companies and all the developers of offshore wind projects (Korsnes 2016). Moreover, project developers in China are large state-owned enterprises (SOEs) with strong financing opportunities and a long history of project development, making them influential in decisions of what projects should be developed (Korsnes 2016).

China has two domestic certification bureaus: China General Certification (CGC) and the China Classification Society. Similar to the foreign companies, these bureaus provide certificates, but they do not have experience with offshore wind projects. They certify wind turbines and turbine components according to the same international standards but their certificates are less acknowledged internationally. So far, they have certified only Chinese turbine designs. In the next section, we crystallise the narratives which each of the identified groups of narrative actors circulated and negotiated.

The role of certification in Chinese offshore wind development

The European certification agency perspective

European certification and advisory agencies (CAA) struggled to gain a foothold in the service industries in China and were acutely aware of the difficulties of getting access to the market. Several of the interviewees pointed out that a big barrier was that the market was protected and that Chinese companies were often privileged. For instance, some project tenders demanded only Chinese project managers, and some tenders were published only in Chinese, which made them hard to discover in time and to forge the right contacts. Another challenge was tied to the fact that employees had to be available to a much higher degree than what would be demanded in Europe. One of the interviewees, a European engineer based in China (Interview #19, Oct. 2013), expressed it in this way:

> Here, I might get an email on Saturday or Sunday from the client and they expect a reply from me. And I don't mind because I understand, but if somebody sends an email to them [his colleagues] in Europe, they might reply with 'No problem, I'll answer first thing on Monday.' But they maybe won't reply until Wednesday, and then the [Chinese] client is [upset] … So, responsiveness is pretty important here.

Thus, availability was crucial in the rapid-growth Chinese environment, where a situation could change literally overnight. Adaptation to the local situation and customs appeared to be a large part of entering companies' struggle to gain access. The entering companies were nonetheless certain that they could offer their services, and particularly so in the offshore wind industry, a European consultant at one CAA (Interview #19, Oct. 2013), stated:

> Generally, they haven't done it [offshore wind projects] before and they don't know how to get from A to B. One of the problems for us is that some of them are very big organisations and you know, they're very, I guess very proud of their organisations … They sometimes believe that they … they don't necessarily need as much help as perhaps we think they do. […] Where they really need us is when they haven't done it before and you see they literally haven't even designed an offshore structure before; so, you know, suddenly we become much more valuable in that case.

This quote underlines the overall finding that the Chinese were perceived as a bit stubborn, wanting to solve things without the help of others. However, the entering company claimed to have experience which could help the Chinese avoid large costs, risks and use of time. This made certification agencies think they should be considered valuable. In fact, the attitude towards local companies was that they were 'arrogant', as a regional manager (Interview #36, Jan. 2014) of one entering company expressed it:

> We have connections with many of the big players, such as [large Chinese developer] and some provincial governments. But [large Chinese developer] are quite arrogant currently in the sense that they think they can do everything on their own. 'We have never done this before, but we are sure we can', you know. I should perhaps not think this way, but it would be good for us if they experience a first big fail, and they realise how important our experience and service is.

Hence, one of the strategies of the entering certification companies appeared to be waiting for the Chinese industry to fail, as there was a firm belief that Chinese companies' lack of attention to risk would, in the end, lead to disaster—disaster which could be avoided, the European companies argued, if their services were welcomed.

Compared to the advisory business, which had a small selection of offshore wind clients, the certification business was the industry which the European companies put the most effort into selling. European companies were quite confident that they would succeed in providing project certificates to local customers. The challenge was identified as the Chinese offshore wind project developers' lack of prioritisation:

> In Germany you need project certification because otherwise you will not get permission to build the wind farm. So, there it is quite obvious

> that you need project certification. In China, they just start [building], and they do not really think about risks. So, project certification is not needed. But anyhow, we want to establish some services, like risk assessment. What we are trying now is to make them aware of some risks and identify some of our services that we can offer separately.
>
> (Interview #24, Dec. 2013)

We see that, although the employees were aware their services not really being wanted in China, they still tried to offer some of the same services they offered in Europe. One strategy was therefore to make the Chinese aware of perceived risks. The need for certification was displaced with a new goal: minimising risk and the extra costs which accidents would create. Moreover, in order to appear less intimidating and to gradually make Chinese customers realise the value of their services, they chose to offer smaller service packages instead of large ones (such as project certificates). The idea was that it was only a question of time before their services would be asked for, as disregarding risks would lead Chinese companies into problems. One of the advisors in a CAA (Interview #25, Dec. 2013) expressed that: 'The Chinese style is just really different to what we do in Europe. In Europe we are very risk averse, but here they are much more, you know, "Just get on and do it!"'. The European company believed that they could make the Chinese more concerned about risk, as managing risk appeared to be crucial for achieving development. Chinese potential customers, on the other hand, appeared to be willing to take more risks and to learn and adapt during the process. These opposing stances of risk-taking in order to learn versus containing risk were also recognised by the European employees of the CAA.

An interview with a European regional manager of a European certification agency made it clear that the company was afraid of losing relevance in China. This was especially so since they were training the local companies in basic offshore wind development techniques, risks and procedures. It was therefore important for the European company to 'make sure we train them in a way so that we are not out of business in some years. Chinese engineers are good engineers, and they learn quickly' (Interview #26, Dec. 2013). A perceived threat was that the Chinese company would acquire enough competencies to render the European company unnecessary. The overall strategy of consultancy and certification companies trying to enter China was naturally to profit from their European experience and to convince potential Chinese customers that their services were sorely needed. However, getting involved early conflicted with the goal of making the Chinese clients understand the risks: if the Chinese were indeed to 'try and fail', the Chinese companies might learn enough to make the European companies obsolete. This was the implicit concession the European companies made by recognising the importance of entering China early.

Hence, the European certification companies were not very successful in soliciting Chinese clients. Their services were not perceived as needed and the strategy of waiting for an accident did not prove effective. Another strategy had to be found, and a third narrative performed by Chinese employees within the European Certification agencies promoted one such strategy.

The perspective of the Chinese offshore wind industry

Chinese offshore wind industry actors appeared to posit a fundamentally different approach to the practices which were argued by European certification agencies. Interviewees in the Chinese industry were quite specific on their ability to develop this industry on their own terms, without much assistance from European firms. They did see a potential for learning from the European experience, but they wanted to perform the actual operations themselves:

> Sometimes I have dinner and drink with the vice president of [a Chinese design company]. I just made a joke that the developers say I can assess some design from your side, and he says 'What, are you kidding? We are the best in China; we don't need your assessment!'
>
> (Chinese employee of a CAA, Interview #26, Dec. 2013)

This statement represents a view put forward by many of the Chinese design companies: they were confident that their competence was on the same level as that of the Western companies, and therefore did not need help despite their lack of experience with offshore wind projects.

On the other hand, some Chinese turbine manufacturers viewed international certificates as symbolic proofs of technology mastery, and therefore as important for learning how to build the technology:

> Certification is taken as a symbol that a manufacturer has established its own capability for the design of a wind turbine. (…) In the early stage people follow the certification process to acquire knowledge; know-how of what the procedure should be like, to learn what kind of procedure they will have to follow.
>
> (Interview #51, Apr. 2014)

As the interviewee pointed out, Chinese companies pay close attention to international standards and certification procedures in order to learn the internationally accepted procedures. However, as certificates were not needed in the domestic market, spending valuable time to gain a full certificate was seen as a waste. As the interviewee (#51, Apr. 2014) clarified: 'A design assessment will do on the market, and it needs less time compared to full type certification. Full type certification is not a must currently'. The Chinese companies only considered getting a certificate from a European company if they wanted to export their turbines.

Conversations with project developers—all large state-owned enterprises—emphasised the difference between Europe and China in their approach to project development. They would go to some of the seminars and workshops of foreign certification firms to get 'inspiration', but their own experience with onshore wind energy was rated more important: They only found a couple of points to be different compared with the onshore wind industry and therefore found it only necessary to 'focus on these very important points, and then it will be no problem' (Interview #53, Apr. 2014). From this I take that the European knowledge was perceived as useful, but not needed—a mind-set widely shared amongst the interviewed Chinese developers. Moreover, conditions in China were different than in Europe. An interviewee from a large project developing company (Interview #54, Apr. 2014) pointed out that there was 'much to learn from abroad in terms of technology', but for the specific intertidal projects 'we had to construct our own vessels'. The intertidal projects were projects constructed along the Chinese coast with no sea cover during low tides. European companies did not have any experience with such conditions, and Chinese developers thus had to develop their own designs. Since the intertidal challenge was overcome with domestic competence, Chinese firms were reluctant to accept advice from European companies with expertise in deepwater offshore wind projects. In other words, Chinese companies did not want help because they did not find European companies competent with their conditions, and because the Chinese were convinced they could do it themselves.

Another interviewee from a large project developing company underlined that local and political challenges was another reason why foreign companies were not able to participate in projects. As he put it: 'for the preconstruction processes, such as policy, planning and approval, it is something we need to sort out here first' (Interview #55, Apr. 2014). In other words, the Chinese saw themselves as knowledgeable about local policy and planning processes and believed that foreign companies would not be able to assist in such processes. This could be a strategy to give domestic players more time to get established, so they ultimately would not need foreign consultants. On the one hand, foreign service suppliers claimed that they needed to be included earlier in the planning process so that risks and costs could be reduced early on. This early phase learning was emphasised as important by most of the interviewed Chinese developers, including a developer consortium of offshore wind projects in the south of China. They had at one point hired some European certification companies for inspections and consultancy but emphasised that 'now in the beginning we have to test and do some experiments on our own in order to learn how to develop an offshore project' (Interview #54, Apr. 2014). Thus, the local Chinese industry underlined the need to have time to do their own experiments with offshore wind project development and was clearly reluctant to let European companies take a role early on. This shows how Chinese companies largely resisted the effort of European companies to access the market and to introduce practices and procedures which had been acquired from European projects.

A middle ground: negotiating risk

All the interviewed European certification companies had a majority of local Chinese employees in their Chinese offices. Throughout the fieldwork it became apparent that the Chinese employees were constantly exposed to the narrative of the European entrant, but at the same time very much in touch with the local Chinese understanding of the situation at hand. Chinese employees of the entering companies therefore took on a mediating role between the entering European certification companies and the domestic Chinese industry. As one Chinese manager expressed it: 'I try to harmonise expectations from both sides, and sometimes the expectations need to be controlled' (Interview #14, Aug. 2013). The Chinese employees appeared to have a good perception of the local Chinese clients' wants and needs, and consequently what they thought about acquiring certification and advice from a foreign company. The general observation was that these kinds of services were not easily sold in China. As expressed by several of the interviewees, consultancy services were not desired because they were seen as 'an unnecessary extra cost' since the companies believed they were capable of fixing most problems themselves. A similar, but yet, different sentiment was voiced by another Chinese manager from a European consultancy company (Interview #52, Apr. 2014):

> It is quite hard to sell services in China, because the Chinese prefer physical things that you can see, and not only what is in your brain. So, we really need to convince people of the necessity of our services.

Hence the technology, *as a physical entity*, counted most in China, and any accompanying *mental activity* was considered something which could be learned quickly. Consequently, it was difficult to sell the consultancy service as such service was first and foremost regarded as an additional cost that was intangible and perceived unnecessary by the Chinese companies.

As we saw in the previous section, the certification companies were concerned with entering early, before local Chinese companies had learned enough to make their services superfluous. The Chinese employees also stressed this:

> If we do not join the first projects or we do not deliver properly, then we may also lose this market totally. You can see this in the construction industry [...] there are no Western companies there. [...] So, we have to give them a similar price to the local companies, easy communication, and localise.
>
> (Chinese business developer, Interview #25, Dec. 2013)

Offering competitive prices and facilitating communication by hiring local employees were strategies employed by the European certification companies to attract clients early on. Interestingly, this interviewee drew a comparison with the construction industry, which had no foreign consultancy; hence, his interpretation was that European companies had not entered early enough. The idea seemed to be that European companies could integrate only by forging a business relationship early on. However, as a Chinese consultant from a European company stated: 'If we get involved in the first stage, we would develop some kind of relationship, and we can always provide some service. [...] So the basis is that we already have a relationship with Chinese offshore [companies]' (Interview #25, Dec. 2013). As the quote shows, Chinese manufacturers were, over time, expected to be able to produce most of their projects alone, without the help of foreign firms. In order to become interesting quickly, the European companies saw the need to develop a relationship with local companies. The strategy was quite unclear, however, as to exactly how such relationships would be established and developed.

One of the topics which created some friction between the European and Chinese employees of the entering certification companies was the suitability of offering project certification in the Chinese market. On the one hand, European employees were quite certain that they should aim to provide the service. Chinese employees, on the other hand, were not so enthusiastic about the idea:

> The first time I had a meeting with [the European headquarters] in 2009 the boss told me, 'you have to develop a project certificate'. My answer was no. At that moment it was quite clear. No market, no driver, no clients. I said we have to start from the type certificate.
>
> (Interview #26, Dec. 2013)

Thus, a type certificate was seen to be easier to sell than a project certificate. This was in 2009, and the European strategy was to provide project certificates for Chinese offshore wind projects. The interviewee was quite confident that this was not the right strategy and offered an alternative approach. A Chinese CAA certification manager chose to rebrand the service by calling it 'risk management', rather than issuance of a project certificate. In other words, he replaced the European practice of certification to a more general concern over risks:

> I tried several times to advertise the project certificate [to Chinese clients] and they were not really interested, because they do not need financing from a bank. But they care about the risk, and they care about the cost. Also, a lot of developers do not really understand what a project certificate is. So, I just repackaged it and do not talk about project certificates. Instead, I refer to it as risk management (...) and developers are really interested in that.
>
> (Interview #26, Dec. 2013)

As this quote shows, risk perception was again at the centre of the story. Chinese clients cared about risk, but they did not pay attention to how project certificates would reduce risk. It appeared that risk, to the Chinese, was connected more to costs, but project certificates did not address this concretely enough. Therefore, the narratives of Chinese employees of European companies were modified so that Chinese interests were accommodated. Chinese employees were more consolidating and less certain about the 'right' approach to the market, acknowledging that the narrative needed to be amended in order to succeed. In other words, they were aware that the strategies needed to be translated in accordance with the visions and needs of Chinese companies and their local practices and understanding of services. Interestingly, over the year spent in Shanghai, the Chinese CAA manager quoted above enrolled more and more into the narrative of the European company, going from being quite certain that project certificates were unsuitable in China to being more convinced that project certificates could be sold there. To the European employees of the certification companies, it appeared impossible that their procedures and standards would not be wanted in China.

Unveiling the intermediaries

European certification companies were relatively unsuccessful in making Chinese clients see the necessity of their services. Some Chinese employees of these companies were indeed trying to reframe the necessity of the European experience in China by referring to potential risks related to developing offshore wind technology. Nevertheless, local industry evaluated the involved risks differently, as they wanted to learn from own experiences.

This chapter has demonstrated how European certification firms failed in selling their offshore wind services to Chinese clients. Project certifications are essential in a European context, so it is clearly surprising to the European companies that their services were not considered important in China. We can learn from this analysis that the Chinese innovation strategy in the offshore wind area and probably other areas as well, is different to that of the European one. This is true on both a governmental level and an industrial level. Several of the procedures which could provide scope for European certification agencies were either taken care of by Chinese government–affiliated institutions, such as banks, developers, construction firms or design institutes, or provided by Chinese certification agencies. Hence, the position of the Chinese government in developing Chinese offshore wind projects demonstrated a notable difference between Europe and China.

On an industry level, Chinese offshore wind companies did not see it necessary to rely on European companies because European companies did not know local policy processes, and Chinese companies believed they were independently capable of understanding the technology and developing domestic procedures and standards. For Chinese companies to gain access to European experience they had to give European actors more control, which could imply less tangible learning for Chinese firms. In this sense, European companies wanted to be seen as an 'obligatory passage point' for Chinese companies to comply with the international standards which had been developed by the same European certification agencies. However, as the analysis shows, they have not succeeded in achieving this position. From a Chinese perspective, the European innovation strategy relating to offshore wind looked like overengineered project development. This stood in contrast to the Chinese innovation strategy wherein taking risks was seen as a precondition for development.

This chapter finds evidence that intermediaries coming from abroad were not very relevant in China's offshore wind industry, and this is also likely to be the case for other Chinese industries. Certification agencies were some of the first European agents to become involved with the learning attempts of Chinese offshore wind firms. However, European certification agencies also attempted to enrol Chinese actors into a global system of certification and standards. Intermediaries may, in other words, also be promoting their own interests, interests which are not necessarily beneficial to innovation processes since they may slow down or hinder learning by Chinese actors.

Instead of accepting intermediaries as 'knowledge brokers', we must ask: Between whom is knowledge shared? A European company might see certification as 'necessary to innovate' whilst a Chinese company understands it as 'ceding the initiative'. With this type of actor-network theory understanding enables a look at the politics behind what some would refer to rather neutral 'facts', 'knowledge' or 'experience'. Instead, risk in a Chinese context relates to more than poorly functioning products or mismanaged projects. For Chinese government and industry actors engaged in offshore wind development, there was also a risk connected to engaging in too much help from abroad—a risk which perhaps was regarded larger than the risk of failing to develop a project the first time around. Therefore, keeping control of the process, learning and gaining own experiences, were more important to Chinese offshore wind actors than managing risks perceived from a European perspective. One implication of this chapter is therefore that there is a conflict between the right design for innovation policy and risk regulation in China, and to resolve this conflict innovation policies and risks must be evaluated locally.

This discussion therefore relates back to what was pointed out by Ang (2016, Kindle loc. 277), namely that starting with what would be termed 'weak' institutions is actually a prerequisite to establish what could be called 'strong' ones: 'the practices and features that defy norms of good governance—normally viewed as "weak" institutions—are paradoxically the raw materials for building markets when none exist'. This has also been well illustrated in the recent book by Kirkegaard (2019), as one of the experts from a foreign research institute put it, talking about China's wind industry development:

> [Quality] was not important in the beginning. The most important was to make the industry try … to let them try out producing [wind] turbines. And then see if any turbines could come out of that, and find out

who were getting good at it, while giving a lot of financial support at the same time.

(Quoted from Kirkegaard 2019, p. 136)

This approach could also be called a strategy of 'letting a hundred flowers bloom', or as Kirkegaard (2019, p. 23) also calls it, the Chinese approach of a 'human wave attack'—a metaphor based on an old Chinese military tactic intending to overrun the enemy line based on large masses of soldiers. Thus, trying and failing based on a large number of attempts was thought to yield results in the end, but not without a high share of failures resulting in 'massive quality problems' (Kirkegaard 2019, p. 23). Today, four years after the interviews with Chinese offshore wind stakeholders were done, it appears that the strategies have changed. One industry specialist interviewed in April 2018 said that the Chinese offshore wind industry is 'much more open to participation from international actors, which is a new thing'. Partnerships with international developers have been made, such as the deal between EDF and China Energy Investment Corporation (Deign 2019), and one of the most common turbines used in Chinese offshore wind projects is a Siemens wind turbine. This implies that the development approach taken in the offshore wind industry has been different from the one in onshore wind industry. In the next chapter, I take a closer look at the way in which Chinese government (central, provincial, local), state-owned and private enterprises and other related actors approached the development on China's offshore wind industry.

Notes

1 Note that private entrepreneurs were able to become members of the CCP earlier but recruiting them was banned in 1989 (Dickson 2007). However, during the 1990s, private entrepreneurs were quietly co-opted by local party officials, and in 2002 the constitution changed so that this informal practice became legitimate (Dickson 2007). According to Dickson (2007, p. 827), this happened for two important reasons: 'first, to seek co-operation between the state and private enterprises, which are responsible for most new growth and job creation, central elements of the CCP's claim to legitimacy; and second, to prevent entrepreneurs becoming an organized opposition'.

2 This section is based on excerpts from a previously published article: Korsnes, M. and Ryghaug, M., 2017. With license to build: Chinese offshore wind firms rejecting European certificates. *Technology Analysis & Strategic Management*, 29(7), pp. 750–761. Material is reused with permission.

References

Andersen, H.W. and Collett, J.P., 1989. *Anchor and Balance. Det norske Veritas 1864–1989.* Det Norske Veritas. J.W. Oslo: Cappelens Forlag.

Ang, Y.Y., 2016. *How China Escaped the Poverty Trap.* Cornell University Press.

BNEF, 2010. Joined at the hip: the US-China clean energy relationship, White Paper, Bloomberg New Energy Finance (BNEF). Available at: www.bnef.com/WhitePapers/download/27 (Accessed 7 August 2011).

Bartley, T., 2011. Certification as a mode of social regulation. In D. Levi-Faur (ed.) *Handbook on the Politics of Regulation*. Cheltenham and Northampton: Edward Elgar Publishing, pp. 441–453.

Bessant, J. and Rush, H., 1995. Building bridges for innovation: the role of consultants in technology transfer. *Research Policy*, 24, 97–114.

Bin, L. 2018. China's solar industry is at a crossroads. Available at: www.chinadialogue.net/article/show/single/en/10775-China-s-solar-industry-is-at-a-crossroads (Accessed 1 May 2019).

Binz, C., Truffer, B., Li, L., Shi, Y. and Lu, Y., 2012. Conceptualizing leapfrogging with spatially coupled innovation systems: The case of onsite wastewater treatment in China. *Technological Forecasting and Social Change*, 79(1), pp. 155–171.

Binz, C. and Anadon, L.D., 2018. Unrelated diversification in latecomer contexts: Emergence of the Chinese solar photovoltaics industry. *Environmental Innovation and Societal Transitions*, 28, pp. 14–34.

Busch, L., 2011. *Standards: Recipes for Reality*. Cambridge, MA: MIT Press.

Chinadaily.com.cn, 2011. New energy exports face trade barriers by Ding Qingfen and Du Juan. Available at: www.chinadaily.com.cn/china/2011-11/17/content_14108416.htm (Accessed 17 November 2011).

CWEA, 2018. Chinese Wind Energy Association (CWEA), Report. Available at: https://community.ieawind.org/HigherLogic/System/DownloadDocumentFile.ashx?DocumentFileKey=46f86d9b-7b90-ab73-f974-e23f97039fb1 (Accessed 01 May 2019).

CWPP 2010. *Wind Power in China 2008: An Analysis of the Status Quo and Perspectives for Development* (2nd ed.). China Wind Power Project and Gesellschaft für Technische Zusammenarbeit (GTZ). Available at: www.cwpc.cn/cwpc/en/system/files/documents/CWPP%20Study_Wind%20Power%20in%20China%202008_Feb2010_ed.pdf (Accessed 27 September 2011).

de la Tour, A., Glachant, M. and Ménière, Y. 2011. Innovation and international technology transfer: The case of the Chinese photovoltaic industry. *Energy Policy*, 39(2), pp. 761–770.

Deign, J., 2019. EDF's China offshore wind deal 'opens door' to western developers. Available at: www.greentechmedia.com/articles/read/edfs-china-offshore-wind-deal-opens-door-to-western-developers (Accessed 11 May 2019).

Delman, J., and Chen, Y., 2008. *Nordic Collaboration with China in Energy Research and Development*. Nordic Institute of Asian Studies (NIAS). Available at: http://copenhagensummeruniversity.ku.dk/en/courses/decodingchina/joergendelman/china_in_energy_rd_nov_08_1_.pdf/ (Accessed 22 February 2012).

Dickson, B.J., 2007. Integrating wealth and power in China: the Communist Party's embrace of the private sector. *The China Quarterly*, 192, pp. 827–854.

DNV 2011. Interview with GDUPC's Mr Xiao Jinsong: Chinese wind power industry expanding globally. *Wind Energy Update*, 2. Available at: www.dnv.com/industry/energy/publications/updates/wind_energy/2011/Windenergy_3_2011/InterviewwithGDUPCsMrXiaoJinsongChinesewindpowerindustryexpandingglobally.asp (Accessed 27 March 2012).

Ernst, D., 2013. Standards, innovation, and latecomer economic development—a conceptual framework. East-West Center Working Paper No. 134. Honolulu, Hawaii.

Garvin, D., 1984. Product quality: An important strategic weapon. *Business Horizons*, 27(April), pp. 40–43.

Green, M.A., 2019. How did solar cells get so cheap?. *Joule*, 3, pp. 631–640.

GTM report, 2011. U.S. Solar energy trade assessment: trade flows and domestic content for solar energy-related goods and services in the United States. A GTM Research Study, U.S. Solar Energy Trade Assessment. Available at: www.seia.org/galleries/pdf/GTM-SEIA_U.S._Solar_Energy_Trade_Balance_2011.pdf (Accessed 7 November 2011).

Gustafsson, I. and Tamm Hallström, K., 2012. Unpacking the certification revolution—the construction of legitimacy, in: Interdisciplinary Perspectives On Accounting (IPA) Conference 2012, 11–13 July 2012. Cardiff, UK.

Gustafsson, I. and Tamm Hallström, K., 2013. The certification paradox: monitoring as a solution and a problem. In M. Reuter, F. Wijkstrom and B. Uggla (eds) *Trust and Organizations. Confidence across Borders*. New York: Palgrave Macmillan, pp. 65–91.

Hargadon, A. and Sutton, R.I. 1997. Technology brokering and innovation in a product development firm. *Administrative Science Quarterly*, 42 (December), 716–749.

Higgins, W. and Tamm Hallström, K., 2007. Standardization, globalization and rationalities of government. *Organization*, 14, 685–704.

Howells, J., 2006. Intermediation and the role of intermediaries in innovation. *Research Policy*, 35(5), pp. 715–728.

Huang, P., Negro, S.O., Hekkert, M.P. and Bi, K., 2016. How China became a leader in solar PV: An innovation system analysis. *Renewable and Sustainable Energy Reviews*, 64 (October), pp. 777–789.

IEC, 2010. Technical committee 88—a background. International Electrotechnical Commission (IEC) [WWW Document]. Available at: www.iec.ch/cgi-bin/getfile.pl/sbp_88.pdf?dir=sbp&format=pdf&type=&file=88.pdf (Accessed 10 August 2014).

IEC, 2019. TC 88. Wind energy generation systems. TC 88 Membership overview. Available at: www.iec.ch/dyn/www/f?p=103:29:15329797158003::::FSP_ORG_ID,FSP_LANG_ID:1282,25 (Accessed 9 May 2019).

Kirkegaard, J.K., 2019. *Wind Power in China. Ambiguous Winds of Change in China's Energy Market*. Abingdon & New York: Routledge.

Koch-Weser, I. and Meick, E., 2015. China's wind and solar sectors: trends in deployment, manufacturing, and energy policy. Report by the U.S.-China Economic and Security Review Commission.

Korsnes, M., 2016. A sustainable Chinese catch-up? Product quality and interactive learning in the offshore wind industry. *International Journal of Technological Learning, Innovation and Development*, 8(2), pp. 172–196.

Lacey, S. and Romm, J., 2011. Will the GOP celebrate the Solyndra bankruptcy to attack clean energy, or will we learn some valuable lessons? Available at: http://thinkprogress.org/romm/2011/09/01/310408/gop-solyndra-bankruptcy/ (Accessed 1 September 2011).

Lema, R., Berger, A., and Schmitz, H., 2013. China's impact on the global wind power industry. *Journal of Current Chinese Affairs*, 42(1), 1, pp. 37–69.

Lewis, J., 2007. Technology acquisition and innovation in the developing world: wind turbine development in China and India. *Studies in Comparative International Development*, 42(3–4), pp. 208–232.

Lewis, J., 2011. Building a national wind turbine industry: experiences from China, India and South Korea. *International Journal of Technology and Globalisation,* 5(3), pp. 281 305.

Lewis, J., 2013. *Green Innovation in China—China's Wind Power Industry and the Global Transition to a Low-Carbon Economy.* New York: Columbia University Press.

Marigo, N., 2007. The Chinese silicon photovoltaic industry and market: a critical review of trends and outlook. *Progress in Photovoltaics: Research and Applications*, 15, pp. 143–162.

Nahm, J., 2017. Exploiting the implementation gap: policy divergence and industrial upgrading in China's wind and solar sectors. *The China Quarterly*, *231*, pp. 705–727.

Nahm, J. and Steinfeld, E.S., 2014. Scale-up nation: China's specialization in innovative manufacturing. *World Development*, 54, pp. 288–300.

nytimes.com, 2011. Trade war in solar takes shape, by Keith Bradsher. Available at: www.nytimes.com/2011/11/10/business/global/us-and-china-on-brink-of-trade-war-over-solar-power-industry.html?pagewanted=all (Accessed 9 November 2011).

Pesendorfer, D., 2011. Risk regulation and precaution. In D. Levi-Faur (ed.) *Handbook on the Politics of Regulation*. Cheltenham and Northampton: Edward Elgar Publishing, pp. 283–294.

Ponte, S. and Cheyns, E., 2013. Voluntary standards, expert knowledge and the governance of sustainability networks. *Global Networks*, 13(4), pp. 459–477.

Power, M., 1996. Making things auditable. *Accounting, Organizations and Society*, 21, pp. 289–315.

Power, M., 1997. *The Audit Society: Rituals of Verification*. Oxford, New York: Oxford University Press.

Power, M., 2007. *Organized Uncertainty: Designing a World of Risk Management.* Oxford, New York: Oxford University Press.

Quitzow, R., 2015. Dynamics of a policy-driven market: The co-evolution of technological innovation systems for solar photovoltaics in China and Germany. *Environmental Innovation and Societal Transitions*, 17 (December), pp. 126–148.

Quitzow, R., Huenteler, J. and Asmussen, H., 2017. Development trajectories in China's wind and solar energy industries: How technology-related differences shape the dynamics of industry localization and catching up. *Journal of Cleaner Production*, 158 (August), pp. 122–133.

Rechargenews.com, 2011. China's wind manufacturers plant seeds of global growth. Available at: www.rechargenews.com/energy/wind/article283563.ece (Accessed 17 October 2011).

Reeves, C. and Bednar, D., 1994. Defining quality: alternatives and implications. *Academy of Management Review*, 19(3), pp. 419–445.

REN21, 2011. *Renewables 2011 Global Status Report*. Renewable Energy Policy Network for the 21st Century (REN21). Available at: www.ren21.net/Portals/97/documents/GSR/GSR2011_Master18.pdf (Accessed 10 August 2011).

REN21, 2018. Renewables 2018 global status report. Renewable Energy Policy Network for the 21st Century (REN21). Available at: www.ren21.net/status-of-renewables/global-status-report/ (Accessed 21 April 2019).

Renewableenergyworld.com, 2011. SolarWorld on trade complaint: 'this was our time to step up' by Steve Leone. Available at: www.renewableenergyworld.com/rea/news/article/2011/10/solarworld-on-trade-complaint-this-was-our-time-to-stand-up (Accessed 20 October 2011).

Russell, S. and Williams, R., 2002. Social shaping of technology: frameworks, findings and implications for policy. In K.H. Sørensen and R. Williams (eds) *Shaping Technology, Guiding Policy: Concepts, Spaces & Tools*, pp. 37–131.

Shubbak, M.H., 2019. The technological system of production and innovation: The case of photovoltaic technology in China. *Research Policy*, 48(4), pp. 993–1015.

Sinovel.com, 2012. Sinovel builds demonstration project in Sweden. Available at: www.sinovel.com/en/newscontent.aspx?ID=389 (Accessed 6 March 2012).

Sousa, R. and Voss, C., 2002. Quality management re-visited: A reflective review and agenda for future research. *Journal of Operations Management*, 20(1), pp. 91–109.

Stewart, J. and Hyysalo, S., 2008. Intermediaries, users and social learning in technological innovation. *International Journal of Innovation Management*, 12(3), pp. 295–325.

Sturdy, A. and Wright, C., 2011. The active client: The boundary-spanning roles of internal consultants as gatekeepers, brokers and partners of their external counterparts. *Management Learning*, 42(5), 485–503.

Tamm Hallström, K., 1996. The production of management standards. *Revue d'économie Industrielle*, 75(1), 61–76.

Tamm Hallström, K., 2000. Organizing the Process of Standardization. In N. Brunsson and B. Jacobsson (eds) *A World of Standards*. Oxford, New York: Oxford University Press, pp. 85–100.

Timmermans, S. and Berg, M., 1997. Standardization in action: achieving local universality through medical protocols. *Social Studies of Science*, 27(2), pp. 273–305.

Timmermans, S. and Epstein, S., 2010. A world of standards but not a standard world: toward a sociology of standards and standardization. *Annual Review of Sociology*, 36 (August), pp. 69–89.

Woebbeking, M., 2008. IEC TS 61400-22 (First Revision of IEC WT 01) [WWW Document]. Available at: www.gl-group.com/pdf/IEC_TS_61400-22_Woeb.pdf (Accessed 10 June 14).

Wright, C., Sturdy, A. and Wylie, N., 2012. Management innovation through standardization: Consultants as standardizers of organizational practice. *Research Policy*, 41(3), pp. 652–662.

Yong, J. and Wilkinson, A., 2002. The long and winding road: The evolution of quality management. *Total Quality Management*, 13(1), pp. 101–121.

Zhang, F. and Gallagher, K.S., 2016. Innovation and technology transfer through global value chains: Evidence from China's PV industry. *Energy Policy*, 94 (July), pp. 191–203.

Zhang, W. and White, S., 2016. Overcoming the liability of newness: Entrepreneurial action and the emergence of China's private solar photovoltaic firms. *Research Policy*, 45(3), pp. 604–617.

Zhou, Y., Pan, M. and Urban, F., 2018. Comparing the international knowledge flow of China's wind and solar photovoltaic (pv) industries: Patent analysis and implications for sustainable development. *Sustainability*, 10(6), no. 1883, pp. 1–34.

6 The future as a resource for change in the offshore wind industry

China's constantly increasing need for electricity, concerns for energy security, wanting to build new industries create jobs for the future, and mitigation of local pollution and climate impact are central reasons why the Chinese Communist Party (CCP) has invested in renewable energy (Ydersbond and Korsnes 2016). In 2009, the Chinese government and relevant stakeholders met to plan the development of China's offshore wind industry (Korsnes 2016). The energy provided by offshore wind was particularly interesting since most Chinese live along the east coast, where wind farms could be built. China's coastal areas are the fastest growing and have, apart from offshore wind, very few energy resources (Naughton 2007). China delved head-on into the offshore wind industry, yet with a troubling start. In 2010, the first concession round for offshore wind farms was initiated; four projects were commissioned in Jiangsu province, totalling one gigawatt (GW) of installed offshore wind power capacity (Zhang *et al.* 2011). In this concession round, all of the developers and turbine manufacturers were Chinese domestic actors. Shortly after, however, the projects were halted and underwent new environmental assessments and cable routing. It was decided that the project sites would be changed, leading to considerable delays and higher costs (Li *et al.* 2012). Nonetheless, based on experiences from the onshore wind industry, many of the largest wind turbine manufacturers began investing in offshore wind technologies.

Industry analysts at the 2012 Offshore Wind China conference registered a pipeline of 37 GW of offshore wind projects in China, suggesting a rather large development compared with the approximately 0.3 GW of offshore wind projects which were installed at the time. Since the first turbine was installed in 2006 in Bohai Bay outside of Beijing, China's offshore wind industry developed more slowly than expected. At the end of 2018, China had a total installed capacity of 4.6 GW (GWEC 2019)—almost reaching the 5-GW capacity target which originally was set for the year 2015. For 2020, the government had set a target of 30 GW, but in 2014 the target was reduced to 10 GW (Smith 2014), and this was again reduced to only 5 GW in the Thirteenth Five-Year Plan (Gosens, Kåberger and Wang 2017).

Government targets started out quite high but were reduced as they were not met, and the industry appeared to not follow up. Figure 6.1 shows the installed capacity of offshore wind in the current three largest markets, occupying 82 percent of the total global installed capacities at the end of 2018 (GWEC 2019). This chapter looks closer at the relation between ambitious government goals and promises on the one hand and China's actual offshore wind industry development on the other.

The future is 'big business' in China, as project developers and government officials attempt to look a bit further over the horizon to be one step ahead (Dodson 2012). At any given time, an industry can gain momentum or remain stagnant, implying that large margins of error are necessary in predictions of the future. This is nicely illustrated through a review of China's *onshore* wind industry during the early 2000s. At that point, it was still unclear whether the industry would take shape. Lewis (2004, p. 84) wrote that

> today with more wind projects cancelled than new projects sited, expectations are less optimistic. National targets for wind power in 2000 were not met, and it is even less likely that China will meet the target of 1,500 MW by 2010.

In hindsight, we know that the expectations of professionals in 2003 proved erroneous only two years later, as China had a total installed capacity of 1,266 MW in 2005 and 2,599 MW in 2006 (Li *et al.* 2007). The comparison with

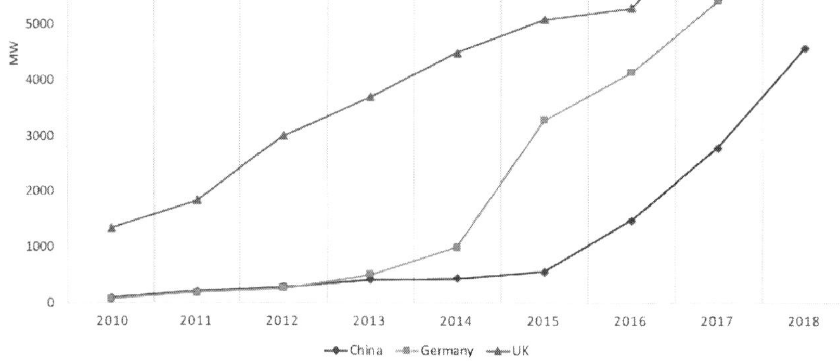

Figure 6.1 Installed capacity of offshore wind in China, Germany and the United Kingdom, in megawatts (MW), 2010–18.

Source: IRENA (2019).

the offshore wind industry between 2010 and 2016 is striking. This retrospection shows that, in spite of the divergence between government targets and other expectations of the future of the wind industry, the industry grew quickly. What role did government targets have in this development, and how did industry actors expect events to develop?

This chapter examines China's offshore wind industry, but one arguably cannot understand this industry without also involving China's onshore wind industry. Therefore, unless otherwise specified, 'wind' industry refers to both the onshore and offshore wind industry. I also make an analytical difference between 'industry' and the 'government' as distinct but interdependent actors, although to be sure, many industry actors in state-owned enterprises (SOEs) also play important governing roles in China, as elaborated on in Chapter 3. Correspondingly, as introduced in Chapter 2, literature on future-orientation and governance is used to denote a difference in expectation levels, namely: 'sociotechnical imaginaries' (Jasanoff and Kim 2009) covering aspects of how governments employ the future as a resource for present-day change, and the 'sociology of expectations', which refers to more local expectation processes connected to a specific technology and technology spokespeople. The role of the government is particularly relevant for two main reasons: First, renewable energy industries are still (but perhaps not for long) dependent on government support globally; second, China has an authoritarian regime in which the role of government is prominent. Hence, we may ask, how does the capacity to imagine futures influence industry development in China? Who has the capacity to influence how the future should look, and how does this process of future-influence unfold?

China's offshore wind industry

By the end of 2013, only two offshore wind projects were completed in China: Donghai Bridge near Shanghai and the Rudong intertidal projects in the intertidal areas of Jiangsu Province (4coffshore 2013). The rest of the installed capacity was one-off test-turbines at different places. China's first major offshore wind farm was the Donghai Bridge 102-MW project and was completed in 2010, in time for the Shanghai Expo. In august 2013, three of the four concession projects that were initiated in 2010 were approved by the NDRC and construction could start (He *et al.* 2016). The three projects were a 300 MW project in Binhai (developed by Datang New Energy), a 200-MW project in Dafeng (developed by Longyuan Power), and a 200-MW project in Dongtai (developed by Luneng Group) (He *et al.* 2016). During 2017, China's wind industry was finally starting to take off. In that year alone, a total of 1.2 GW of installed capacity were added spread across 18 different projects (GWEC 2018a). Of these, almost one GW of projects was located in Jiangsu Province (GWEC 2018a). Today, expectations to the industry are once again high, and experts are expecting that China's offshore wind market will grow quickly, overtaking the European offshore wind market already in 2021 (Deign 2019). As an industry specialist interviewed in April 2019 (Interview #77) explained:

> Offshore wind is now actually happening. 2018 was a big year and the industry really kick-started. Developers have gotten more accustomed to the industry, and several new constellations have emerged. In 2018 alone, 16 GW of new projects were approved, and 7 GW are under construction.

We thus see that the industry has gone from high expectations around 2009 and 2010, through a period of little activity and lower expectations and into a new period today, where it appears the industry is starting to take off. This chapter explores this fluctuating period of changing expectations, and what it meant for the development of the wind industry.

China's total exploitable offshore wind resource is estimated to be around 200 GW (at water depths of 5–25 metres), with an additional potential of 500 GW at deeper water depths (50–70 metres) (He *et al.* 2016). According to a report prepared by the Chinese Wind Energy Association, the strongest winds are in the Taiwan Strait off the coast of Fujian, followed by Zhejiang and the west coast of Hainan (WWF 2010). Guangdong also has a significant potential for harnessing winds offshore. However, over the past 50 years, 'Guangdong has had the highest occurrence of typhoons with approximately 160 typhoons coming on land' (WWF 2010, p. viii). The adverse effect that typhoons have on wind farms is a technical challenge which needs to be overcome in order to develop offshore wind projects in typhoon-prone provinces. The occurrence of typhoons makes Jiangsu Province more attractive for development at the present stage, although the project pipeline for Guangdong is high. According to a recent analysis by Azure International, as much as 8 GW of offshore wind projects will have been developed and fully connected in Guangdong by 2021, and for Jiangsu Province, the estimate is about 11 GW—making these two provinces the most important ones in terms of near-term development (Beaumont 2019). The 2021 landmark is important because in May 2018, the NEA introduced a new policy saying that projects approved before the end of 2018 and in operation before the end of 2021 will receive a FiT of 0.85 RMB/kWh (Yu 2019). This was part of an effort to reduce government subsidies and make market-mechanisms more decisive for project approval (Yu 2019). Therefore, for all projects approved after 2018, prices will be decided by tenders and bidding (Yu 2019). More than 40 GW of offshore wind projects were approved before 2018, but the expectation is that only about half of these will actually be in operation before the end of 2021 (Beaumont 2019). Several provinces in their current five-year-plan periods have established development plans for offshore wind. This includes Hebei, Shandong, Jiangsu, Shanghai, Fujian and Guangdong. In Jiangsu Province the 2020 installed offshore wind power target is 3.5 GW, in Guangdong Province, 2 GW and in 2017 the National Energy Administration (NEA) approved a 13.3-GW development plan of Fujian Province (CREO 2018; GWEC 2018b).

In 2015, there were ten companies which produced operational offshore wind turbines in China (Korsnes 2016). Three of the companies were part of centrally owned SOEs, and five were part of SOEs. Two of the manufacturers, Mingyang and Envision, were privately owned, and two companies, Goldwind and Envision, were purely wind power manufacturers (Korsnes 2016). Sinovel, Shanghai Electric, Dongfang, XEMC Wind, SANY and Mingyang are subsidiaries of large industry groups which had their main industries within power generation equipment or heavy machinery. United Power is a subsidiary of the world's largest power generation company and developer, China Energy Investment Corporation (until 2017 Guodian, mother company of Longyuan), and the last company, CSIC Haizhuang, is a subsidiary of the large shipbuilding group China Shipbuilding Industry Corporation (CSIC), a group which also produces gearboxes and other key wind turbine components. In 2017, the market leaders in terms of offshore wind turbines were Shanghai Electric (50 percent), followed by Goldwind (18 percent), Envision (17 percent) and CSIC (9 percent) (GWEC 2018b). Sinovel, which used to be the leading Chinese offshore wind company, has not been visibly active since 2012.

Similar to the onshore wind industry, entering joint ventures and licensing technology from Western firms were frequent strategies of acquiring technology in China's offshore wind industry. Korsnes (2016) found that being pressed for time and money complicated processes to ensure product quality and offshore wind turbine reliance. Time constraints were also identified as an important reason why certification was seen mostly as a symbolic and impractical process. Therefore, in the offshore wind industry, quality conformity was found to be unimportant in terms of procedures, but important for 'face value', which accelerated the commercialisation of the turbines (Korsnes 2016, p. 185). Chinese offshore wind companies faced several challenges related to quality as they attempted to learn to master offshore wind technologies (Korsnes 2016). Health, safety and environment issues were also found to be pertinent under-prioritised issues in the industry. Chinese offshore wind firms saw the European standard as a quality benchmark, and turbine manufacturers opted for 'quasi-certificates' with a foreign stamp which was satisfactory for the project developers and gave them an edge on the domestic market—because developers did not see the difference between the types of certificates. An interviewee from a Chinese certification agency pointed this out (Interview #51, Apr. 2014):

> Some of the [turbine manufacturers] will seek a design evaluation or a prototype certification as defined in IEC standards. These evaluations look quite like the certificate and most of the developers don't recognise the differences between certificates and a design evaluation or a prototype certification. And a design evaluation is much quicker than a full certificate.

In addition, some of the challenges were found to be connected to the ties between suppliers and manufacturers in the supply chain, especially when ties were either too close or too shallow (Korsnes 2016). As an example of the former, the relationship between the large SOE Longyuan and its subsidiary turbine manufacturer Guodian United Power was so closely tied that feedback between supplier and demander was not used to improve product performance: Guodian United Power was the untalented son which received unrelenting support. On the other end of the spectrum, shallow ties led to 'supplier shopping' and communication blocks, leaving scant opportunity for improving and optimising components of the end product. The advantages of close and shallow ties are that local companies are chosen. However, learning opportunities were missed in both cases: when ties were too close, there was a sense of duty or friendship leading to too much patience, and when ties were too shallow, supplier shopping or arrogance was characterised by a lack of patience. Better feedback between supplier and demander could enhance learning processes and, ultimately, product quality. By breaking down the way in which close and shallow ties influenced learning within SOEs and enterprises, more generally, in China, Korsnes (2016) showed how state ownership influenced the acquisition of technological capability. Companies could invest all the time, money and engineers they wanted on research and development, but considerations related to supplier-manufacturer feedback could still hamper learning processes.

In this way, protection of domestic state-owned firms did not unambiguously help the development of China's offshore wind industry. In other words, the mobilisation of large amounts of resources and alliances is not sufficient if certain aspects of the technology are not mastered, because the technology may not work as intended. Although the role of SOEs is unfortunate in some respects, there are several aspects which make SOEs important, as we shall see below. Today, several Chinese manufacturers have developed their offshore wind designs with certificates from the CGC, and in 2017, manufacturers such as Goldwind, Envision, Mingyang, United Power, CSIC Haizhuang, and Shanghai Electric released new offshore wind designs. Goldwind released a 6-MW offshore wind turbine, Envision released a 4.5-MW turbine, and Mingyang released a 5.5-MW semi-direct drive typhoon resistant offshore wind turbine in 2017 (CWEA 2018).

Ambition and ambiguity: expectations and imaginaries[1]

To refresh the discussion from Chapter 2, expectations as understood in this book are connected more 'locally' to a specific technological project, whereas sociotechnical imaginaries are long-term, state-induced desirable futures. Expectations can work to break new ground, create protected spaces, be self-fulfilling and mobilise new actors to support or oppose a technology. We can assume that collective expectations are malleable and can be drawn upon by governments. Expectations can also be contradictory, and we can expect the Chinese government and industry stakeholders to attempt to make certain expectations more credible than others. These expectations, in turn, impact the government-supported imaginaries. We can expect sociotechnical imaginaries to be present as general understandings of good and desirable futures in the social world writ large. As we turn to look closer at China's offshore wind industry, we may ask what the important elements of successful visions are in this context. What is the relation between offshore wind industry expectations and imaginaries in China, and how do the different actors' perspectives on the future impact offshore wind development?

In a Chinese context, sociotechnical imaginaries associated with wind power can be understood as part of a larger process of nation-building and plan-based governance. Since 1978, economic modernisation has been central to China's 'open-door' policy shift, and market forces have increasingly been allowed to influence the economy (Saich 2011). Although China has a history of heavy government involvement and control under socialist planning, China has been able to change the governing regime by retaining certain governing mechanisms, such as development plans, and eliminating other more detailed control-mechanisms (Naughton 2007; 2010). Chinese sociotechnical imaginaries are typically present in Five-Year Plans, medium- to long-term plans and Special Program Plans (Heilmann 2010), but also surface in school curricula, media and other official government statements. Examples of prevailing sociotechnical imaginaries since 1978 can be 'economic modernisation', the 'moderately prosperous society', 'harmonious society', 'innovation-orientated nation', or the more recent 'Chinese Dream' and 'Ecological Civilisation' proclaimed by President Xi Jinping on a number of occasions since 2012 (Brown and Bērziņa-Čerenkova 2018; Geall and Ely 2018; Heilmann, 2010; Saich, 2011). These imaginaries generally inform and influence decision makers and the general public alike, and they all share an aim of mobilising science and technology in a project of ultimately making China a developed nation. Indeed, this has also been confirmed in *China Daily*, an English-language CCP-controlled newspaper:

> [Ecological civilization] is not a term the Party has coined just to fill a theoretical vacancy in its socialism with Chinese characteristics, but rather a future-oriented guiding principle based on the perception of the extremely high price we have paid for our economic miracle.
>
> (China Daily 2007, cited in Geall and Ely 2018, p. 11)

In recent years, environmental sustainability has gained increasing space in such imaginaries (Economy 2010; Li and Wang 2012; Saich 2011; Stensdal 2014), and wind industry development has gained traction because of motivations such as sustainability and climate change (Lewis 2013; Ydersbond and Korsnes 2016). Such imaginaries are therefore relevant, and we now turn to see how they have played out in the offshore wind industry.

How offshore wind expectations were generated

During my 2013–14 fieldwork in China, it became apparent that the way in which people talked about offshore wind power was anchored in two main experiences: the existing *European* offshore wind industry and the domestic *onshore* wind industry. These two experiences were typically referred to when assessing the speed and quality of China's offshore wind industry progress. Europe frequently served as a basis for comparison regarding the quality of performance in China, which was evident, for instance, in a discussion of the tendering process for offshore wind projects with a Chinese employee of a European certification company (Interview #14, Aug. 2013):

> The Chinese tenders are not serious. In Europe, everything is prepared well, before a bid is made. But if you do that here, you lose! Sometimes the winner is already decided by the government. You can tell this if you read the tender documents; you can already know who they want to win the tender.

In this case, it is implied that the tender which was held for offshore wind projects in 2010 lacked sufficient preparatory work to ensure the feasibility of the projects, which was apparent even to the developers bidding on the projects. In fact, a representative of a large developer (Interview #39, Mar. 2014) justified the low bid prices and, hence, the low probability of starting the project by saying: 'If you jump from the 25th and the 20th floor it's the same, you are dead anyways'. As the previous chapter showed, in Europe, these tenders would have been organised differently, often supervised by an advisory agency to ensure that potential pitfalls were accounted for.

When referring to the *speed* of offshore wind development in China, interviewees particularly looked to the development of the onshore wind industry. On the one hand, some interviewees hoped that the development of the offshore wind industry would not be as fast as that of the onshore wind industry; on the other hand, they were expecting it to develop with the same swiftness. A Chinese interviewee of the European certification company (Interview #26, Dec. 2013) compared the development of the offshore and onshore wind industries in China, stating:

> I hope the government will take it slowly this time, as opposed to what happened in the onshore wind development, which developed far too fast. It was five, six years of dramatically rapid growth without control. And all the quality issues and everything happened because of that. So, I believe the government needs to learn from onshore. The problem here in China is that when the top decides, it will have to happen. And they in turn put pressure down to the developer, and to the industry. That's a kind of dilemma. Personally, I would think a time-horizon of 20 to 30 years would be good [for the industry development]. This way I can keep my job for a longer time [laughter].

We see that the onshore industry was used as a base for comparison, with the hope that this time around, in the offshore wind industry, it would be better to take it slow. This concern over developing the offshore wind industry too quickly recurred in several interviews. For instance, an engineer from a large Chinese wind turbine manufacturer (Interview #13, Jul. 2013) believed the following: 'The government has observed some problems from the onshore industry, and now they will be more careful. The onshore industry was really booming, but the government has learned, step-by-step'. Again, we see that the government was perceived as the main actor for deciding the speed at which the industry would develop, and industry actors merely followed the tune of the government. Interviewees typically referred to concrete examples of companies which had developed too quickly, had over-expanded and were subsequently on the verge of bankruptcy. One such company was Sinovel, at the time the largest onshore and offshore wind turbine manufacturer. As a Chinese manager of a European wind turbine manufacturer based in China (Interview #47, Apr. 2014) stated, 'Sinovel only focused on expanding their capacity, expanding their market share and did not think about quality that much. Today, the situation is really bad there, and most of their employees are really frustrated'. Interviewees hoped that a greater focus on quality would be prioritised in the offshore wind industry. Hence, industry stakeholders constructed their own likely futures of offshore wind development based on their awareness of the pitfalls which had occurred in the onshore industry.

Some interviewees did not necessarily view the rapid development of the onshore industry as a learning experience which would slow the development of offshore wind energy. Rather, they assumed that the offshore wind industry would follow the same growth pattern. A spokesperson from a large developer said that they 'have an impression recently that the government wants to develop this industry fast. And we can use onshore as an example: after the onshore feed-in tariff was set the industry developed quite fast' (Interview #53, Apr. 2014). There were also other examples of people expecting the development to happen quickly, as had occurred in the onshore industry. An interviewee from a coating manufacturer (Interview #46, Apr. 2014) expected things to 'go crazy' because of what had happened with the onshore industry:

INTERVIEWER: How fast do you think the offshore wind industry will develop in the next five or ten years?

INTERVIEWEE: I don't know. It's usually quite crazy here. When it starts it just goes crazy. You know during 2011 we were really astonished about the speed of the onshore wind development. The amount of new projects … it was just tremendous.

Another interviewee from a wind farm developer (Interview #39, Mar. 2014) thought that the offshore wind industry would 'copy the mode of the onshore industry'. He added that, over the next couple of years, many companies would enter the industry: 'when everything goes well everybody want to join together'. For these reasons, we may conclude that onshore wind experiences calibrated expectations for the offshore industry, in the sense that the growth could be very quick, and, on the other hand, that rapid growth could cause serious costs in terms of low performance, which had been well documented in relation to the onshore wind industry (e.g., Gosens and Lu 2013; 2014).

Other interviewees were generally sceptical that the industry would develop quickly, simply because there were still too many obstacles for offshore projects and onshore resources were still abundant. In this manner, the onshore wind experience became a competing expectation for the immediate future. As a central government policymaker (Interview #30, Dec. 2013) put it:

INTERVIEWER: How do you think the offshore wind industry will develop in the future?

INTERVIEWEE: I don't think it will be quick. So far it is inland, and not offshore wind (that has been fast). Because there are many, many things that should be done if you go to offshore wind. Seabed, resource assessments, and also dealing with the different groups, such as the fishing industry, the ocean administration, and other coordinators. Yes, many things. So, I don't think it will happen in the future three to five years. However, for onshore, we already have done a lot of detailed work.

Today, about five years later, it appears that this interviewee was relatively close in his estimates. Other interviewees expressed concurring views, and low wind-speed areas onshore appeared to be one reason contributing to the delay. An analyst from a certification agency (Interview #51, Apr. 2014) put it this way:

> I know from the National Energy Administration that they would like to have more wind energy in the south. I mean, what's left to build out onshore has lower wind speeds than offshore, and for maybe the next two or three years people will still focus on these wind areas and not on offshore.

A professor from a high-standing university conducting research on offshore wind in China also pointed out that 'right now the most valuable resource is the low-speed wind onshore resource, and this is where the government and the industry are focusing right now' (Interview #23, Dec. 2013). However, the low wind-speed areas were not the only competitors to offshore projects. Local governments, consistently concerned about economic growth in their areas, had other potential future uses for their coastlines. For example, a manager from a foundation manufacturer (Interview #42, Mar. 2014) explained that the local government around Binhai, where one of the concession projects was located, was against the offshore wind projects:

> The Binhai new district expected to develop their shipping business there, and the accompanying logistics, and real estate, and also fishing. So

> obviously the offshore wind development will contribute negatively to this kind of plans.

As we can see from this quote, the various actors envisaged different futures. The future thus demonstrated interpretive flexibility which would bring intended futures more in line with the interests of the spokesperson, and negotiations were constantly ongoing in an attempt to determine whose future was more important.

The comparison with Europe was also used as an argument for onshore wind resources. In Europe, it was claimed, offshore wind was necessary because of land constraints—which was not the case in China. This argument, for instance, was made by a central government policymaker (Interview #28, Dec. 2013):

> This is the different situation here with Denmark, Germany and the UK: In Europe the inland almost does not have any more space for wind, but that's different in China. In the future we must use the ocean, but for now we still have some work to do [onshore].

The development of offshore wind was therefore not considered pressing, and according to the above interviewee's own estimates, offshore wind would become more important in perhaps 'five or even ten years'. However, another interviewee from a large developer (Interview #53, Apr. 2014) said that the reason offshore wind energy was preferable was precisely *because* there was too much land in China and the distance between people was too great:

> In China it's not like we don't have land anymore, like in Europe. I mean, in Inner Mongolia there is much land, but the problem is that the distances are too great there to the load centres, after the electricity is generated. In the southeast of China, like the Jiangsu, Zhejiang and Fujian, they are very industrially developed areas, and when we generate electricity there it is much easier to transport it. And also, the wind resource offshore is better than onshore. So, after some testing we think that we can produce some profit from the offshore wind industry.

Here we again observe that the same conditions were used in two diverging arguments. Above, we observed how the onshore wind industry created diverging expectations for the offshore industry. In this case we on the one hand see that Europe's *lack* of land was viewed as the reason why offshore wind became important there, and on the other hand, China's *abundance* of land (i.e., the great distances between the onshore wind resource and the load centres) was the reason why offshore wind was deemed necessary. Although it appears that the industry stakeholders did not completely agree on the importance of either the onshore industry or the European industry for China's offshore wind industry, we may conclude that they drew heavily on previous experiences in these realms to adjust their expectations of the future. Indeed, actors drew on examples which supported their construction of the future: Different actors had diverging intentions of how the future should look. This interpretive flexibility made people draw on experiences depending on their future interest. As we have seen, some wanted to keep their job longer, others wanted to make sure that they did not lose face and still others focused on ensuring high-quality products. The bottom line is that the different stakeholders used prior experience to tune their expectations in a desired direction according to their situation and the likelihood of achievement. The effects of these expectations therefore depended on the relation between the future and the actor 'reading' the future in question. In this sense, the sociotechnical imaginary was also affected, and the government would adjust development targets. This relation between sociotechnical imaginaries, expectations and actual industry development is examined further in the next section.

The interrelation between expectations and sociotechnical imaginaries

The ambitious government targets for the development of China's offshore wind industry appeared to be strong indicators of the future for industry participants. If an industry receives attention from the central government, the legitimacy and status of the industry is elevated on a provincial and local level, and the likelihood of its success increases (Kong 2009; Korsnes 2014; Shi, Markóczy and Stan 2014). Government targets were also important for attracting new investors. A common perception in the industry was described, in 2011, by an interviewee from a wind energy advisory organisation based in Beijing (Interview #1, Sept. 2011): 'If I am an investor and I want to invest either in solar, biomass or wind in China, I would go to wind because it has a really high development target'. Chinese government officials tended to look many years into the future when formulating their vision, as explained by a central government official (Interview #30, Dec. 2013):

> We focus on the future, to 2050, so you know all these pollution issues you cannot change in one day, or one year. You must take a long time. You should be clear that by 2050, in 40 or 50 years, what is your challenge, your situation? So, if we are clear about the future, everybody can reach a common idea. Now, we come back, from now we should do some work. We must do some work. Otherwise, the future will not wait to come.

We see quite clearly here how the future was enacted in the present and how the government was convinced that ambitious action was required 'today'. These considerations were manifested in the sociotechnical imaginary envisaged by the government, and as the quote indicates, they were motivated by concerns of 'pollution issues'. Hence, government targets may in this case be understood as rather concrete manifestations of a wind industry sociotechnical imaginary, a tool which would help the government reach their desired future.

Although *official* government development targets for offshore wind were high, interviews with policymakers and government agencies indicated that their own expectations with respect to reaching the 2015 and 2020 targets were lower. One policy advisor noted that the 2015 development target had been changed from five GW of 'installed' capacity to five GW of projects 'under construction', reflecting the acknowledgement of central government that the goals may not be reached. The new wording provided a much broader time span for the actual completion of projects, as the phrase 'under construction' was not defined and a project could therefore potentially continue for several years into the future. Hence, the government's ambitious prospects created the impression that more was happening and thus allowed government officials to save face regarding the official development targets. Moreover, the ambitious targets made the industry believe that things looked promising and therefore generated more interest.

The Chinese government overestimated the future, which might have been a strategic decision given that government officials were less ambitious when talking in person. However, industry participants were also aware of the government's overestimations. In effect, offshore wind industry participants in China referred to the industry as an unrealised idea which was likely to happen. As one manager of a certification company (Interview #26, Dec. 2013) put it: 'Currently the Chinese offshore industry is like a cake. So far, it's a really beautiful picture, but there is no real cake here. So, everybody thinks, "I have to be involved or can take some share of it"'. The high expectations made the Chinese industry an attractive cake from which many wanted a slice. Although most industry actors agreed that the 2015 target would not be reached, they were certain that the industry would take off at one point, and at that time, they needed to be well-positioned to get a share of the cake. Government targets can be considered constitutive of sociotechnical imaginaries, thus explaining their ambitiousness, but these targets were also depending on and shaped by local assessments of the future. Futures were constructed based not only on the government's high promises but also on the local expectations of industry participants.

Interviews with various offshore wind industry stakeholders showed that their expectations were much lower than the official development target. Industry participants were generally struggling to gain an overview of the situation at hand, constantly assessing what had actually been done by other actors, what was under construction, what orders had been made, what was only at the planning stage, what the 'tone' was from the central and local government officials, and so on. One Chinese consultant at a European certification company (Interview #25, Dec. 2013) emphasised that it was important to constantly communicate with the various stakeholders to manage their expectations:

> All the international companies have the same questions. For example, you receive the list from the offshore event organisation, that some projects will be kicked off next year. And at the workshop I talked with the

> director of [a design company] the whole evening. We discussed which project can be kicked off, and also which developer. We really went into detail. Each and every project has its reasons why they can kick off or not. Of course, three years or maybe only three months later this will change. So basically, you know that you should not really think that 'Oh it's really good, we have ten projects to be kicked off and we can make our budget based on that!' There will be trouble for everyone if that happens.

This quote shows that the future professed by the government was checked and balanced against the reality of other industry participants, who felt it was important for them not to base their expectations on incorrect assumptions. Because it was difficult to know what and when things would happen, industry participants needed to exchange information. The list mentioned in the quote above was published by the government and outlined the projects which would be launched at what time. However, the timeline described in the list was much more ambitious than what actually occurred.

As reflected in the interviews and industry reports, the most common reason why things did not develop as quickly as expected in the offshore wind industry was the disagreement between the departments responsible for the development: the National Energy Agency (NEA) and the State Oceanic Administration (SOA) (Quartz+Co 2013; Korsnes 2016). This disagreement resided mainly in the procedures of project approval and project siting. Indeed, the necessary governing mechanisms were not in place to accommodate the government targets. As a turbine manufacturer said in 2013 (Interview #21, Nov. 2013):

> Two years ago, the government themselves didn't know how to approve the offshore projects. They didn't know how to do environment assessment for offshore, or what kinds of feasibility studies were needed. I have to say they are still learning.

A project developer (Interview #39, Mar. 2014) agreed with this statement, adding that 'the procedures were not in place to support this kind of projects, so they had to discuss this with other national departments to develop the policies'. The central government needed to establish the procedures for project development and looking further into the future and setting ambitious targets allowed them to quickly mobilise and coordinate actors internally (within the government) as well as externally (in the industry). Thus, as was pointed out in Chapter 4, the government approach was also part of a policy-learning process which went hand in hand with the technology-learning process.

Most actors were waiting for the government to introduce an acceptable electricity price (feed-in tariff) so that offshore wind projects would pay off (Yang 2014). Until 2015, all of the developers of offshore wind projects were Chinese state-owned companies (Korsnes 2016) and were therefore motivated to reach the development targets. However, as they did not receive funding for this, they had to use money from their internal budget, as explained by a manager from one of the largest developers (Interview #53, Apr. 2014):

> We cooperate with the NEA [National Energy Administration], but actually we are the ones who pay for [a demo project close to Shanghai]. We proposed the project to the NEA, and they approved it, so in terms of technology support, opportunities and policies we got many benefits from the government.

In other words, the NEA supported the company in a nonfinancial way, but the company had to pay for the actual development. In this manner, the government was able to ensure that prices were not unrealistically high, and that the central government would not end up subsidising the industry more than necessary. Moreover, it was clear that the government depended on the industry to make the initiative and start development. The industry actors were also aware of this reasoning and used it to gain increased support from the government when needed. This leverage was particularly clear with the four concession projects, totalling 1,000 MW, that were initiated in 2010 but were stalled for several years. An official from the Jiangsu provincial government renewable energy industry association (Interview #37, Feb. 2014) explained the situation:

> The reasons the government gives to the public of why the projects have been postponed is because of the disagreements between government agencies. However, those are just public reasons. The real reasons, actually, I have met with all the related companies for these four concession projects last week, and they say that the main reason is because of the technology. They do not think the technology is mature yet, and the cost is very high.

In other words, the developers did not start constructing the projects because they were too expensive, and they were waiting for the government to increase subsidies and support.

Some developers were also waiting for approval from the right holds before they could start the projects. This was pointed out by a developer in southern China (Interview #54, Apr. 2014), who was planning a 300-MW project:

> We approach this project step-by-step, as we need to get all the necessary approvals from the different agencies. After approval we can start the construction, but right now we are just waiting for approval. Nothing works without approval, but we think we will get it soon.

However, who is the 'government' in this respect? In fact, few appeared to know exactly who the decisive actors were. This uncertainty with respect to the government was prevalent in much of the interview feedback. For instance, one interviewee from an industry association (Interview #4, Sept. 2011), which advised the government of renewable energy policies to be implemented, answered that 'no one really knows these things' when asked at what time she believed the top-level government would start developing offshore wind projects. Only a few select individuals knew about the 'top-level' plans. A foreign energy consultant who had been working with wind energy in China since 2002 (Interview #10, Nov. 2011), made a similar point when asked how policy had been set for the onshore wind industry in China:

> A lot of these things are run by the ERI [Energy Research Institute] and people in the wind energy association, or the renewable energy association. These people often have dual roles, they're either professors or industry association people and they are relied on to advice the NEA [National Energy Administration] and write draft policies. The problem is you never know who has the final word. I mean the academics draft the stuff and then politically it is expedient. But does it happen or does some big guy just walk in the room and say, 'No, it has got to be this way'? Nobody knows.

Industry actors were therefore in the dark as to exactly what would happen and when. Everyone apparently was prepared to 'jump on the train', as the government had announced ambitious development targets, but few knew exactly when the train would depart and in what direction it would head. Thus, the government remained obscure, and industry actors had the impression that things could change at any moment. This uncertainty affected expectations; people needed to be open to both rapid change and no change. Small offices could grow large literally overnight but could also stay small. The government remained ambiguous because it needed to sort out the internal organisation of projects. However, the mysteriousness kept industry actors alert and ready to make a move when the signal was right. Industry participants needed patience, and they needed to expect that something would happen at one point in the future. The relationship between government and industry became a strategic waiting game—a waiting game which was characterised by ambitious government intentions but ambiguous signals in terms of implementation.

The quintessence of the strategic games was rooted in conflicting expectations: The government could, at any point, introduce a feed-in tariff or approve projects and companies who wanted a piece of the cake had to be prepared to seize the opportunity. It was, however, not a waiting game of 'wait and see' but rather 'wait and prepare'. Interviewees from Chinese industry appeared to have a reflexive attitude to government imaginaries: They knew that something would happen, just not when. The performativity of expectations was not located in the circulation of expectations as such but rather in expectations about the performativity of imaginaries. Following this line of reasoning, we arrive at a new interpretation of the ambitious government targets. The government depended on industry input to develop procedures, assess costs and get things done. Industry actors, however, did not accept the validity of the vision in its entirety and were able to convey their version of what was possible and affordable. The government then modified its targets accordingly, for instance, by changing 'installed capacity' to 'under construction' or by actually reducing the development targets. The relation between sociotechnical imaginaries and industry expectations as defined in this chapter could thus be considered a planning tool—a litmus test of what was feasible—that the Chinese government could take strategic advantage of to achieve intended futures. Ambitions shaped the road ahead and ambiguity became a space for negotiation—a reality check for plan-fixated technocrats and cost-concerned companies. In this way, futures became a resource which the CCP could draw on to achieve desired change.

Ambitious yet ambiguous futures

This chapter has examined the role of sociotechnical imaginaries and expectations in China's offshore wind industry to better understand the role they play in industry development in a Chinese context. In particular, we have looked at how sociotechnical imaginaries were negotiated through local expectations, how expectations were generated from experiences with the onshore wind industry and Europe's offshore wind industry, and finally how industry development was characterised by what was called 'strategic waiting games'. The chapter first described the manner in which government targets were set and later negotiated between the government and industry. Given that the early developments of the onshore wind industry continually missed development targets, there is a fairly strong indication that development targets were set rather ambitiously in China. This ambitiousness generated interest in the industry and government, pushing policymakers to move the issue up the agenda. In this manner, a promise and requirement cycle, as van Lente (2000) describes, was generated, and industry participants expected the government to follow through with support schemes for developing the industry.

In general, the basic tenets of the sociology of expectations appear to resonate well with the Chinese offshore wind industry with respect to promise and requirement cycles and strategic games between technology spokespeople and others. However, the interaction between government-induced sociotechnical imaginaries and industry expectations provide new perspectives on how nation-specific and industry-specific expectations can merge into quite concrete actions and plans. By describing a strategic waiting game between the government and industry actors, this analysis has emphasised the fact that Chinese industry actors had to be prepared for both change and inertia. In China, companies and politicians were future-orientated and looked at what the various long-term plans had in store for them. Things could change quickly from one day to the next, and companies therefore had to be prepared to be taken by surprise.

I identified two characteristic components of the waiting game: First, companies delegated responsibility for their future activity to 'the government' and waited for its call. Second, the government remained obscure, leaving industry participants on their toes about what would happen next. More importantly, however, keeping the future ambitious but the implementation ambiguous appeared to be a strategy employed by the government to ensure change without needing to detail the road ahead. Thus, the dynamic relation between the government-induced sociotechnical imaginary and industry expectations functioned as an important planning tool which helped calibrate industry development.

The chapter has established that the government played an important role in constructing successful visions in China. By describing a nuance between expectations and imaginaries, this chapter has contributed to a better understanding of the dynamics of future-orientation in China: As industry actors typically expected something to happen, the negotiation centred on *the extent to which* government imaginaries could and should be performative. Expectations in China materialised more concretely than indicated by both the sociotechnical imaginaries approach (Jasanoff and Kim 2009) and the sociology of expectations (Borup *et al.* 2006). With a plan-based economic history, China's visions were bordering on loose plans, and the theoretical distinction made here between imaginaries and expectations help us understand better how the government was able to make use of conflicting understandings of the future to attain more realistic plans. Government targets which 'failed' to be reached were therefore not necessarily failures per se, but indicators of what industry actors could tolerate in terms of costs and development speed, which in turn were used to set new and more realistic targets.

In practical terms, the described dynamics of future-orientation in China have several implications. For instance, the waiting game appeared to be biased towards larger companies which had the means to keep large capacities idle over time and could improvise and jump on the train when the right government signals were given. In China, these companies typically included the large state-owned enterprises. The waiting game could therefore be used strategically to ensure that domestic (state-owned) companies gained a larger market share than did other (private or foreign) companies. In effect, the waiting game can be considered a protective mechanism which may have given domestic companies a buffer zone to help them acquire the technologies of frontier companies. Another implication is that, as these concrete expectations were negotiated and manifested, the initial imaginary set by the government had a big impact in terms of the direction in which the industry developed in China. Had the CCP not wished to develop renewable energy technologies in the first place, not much would have happened. In a sense, the solidity of expectations in China rested on the legitimacy of previously fulfilled expectations in other industries. Ironically, therefore, the prior success of Chinese companies in developing coal power industries, for example, may be a reason why expectations and ambitions are so high for China's offshore wind industry today.

Note

1 This section is based on excerpts from a previously published article: Reprinted from *Technological Forecasting and Social Change*, vol. 107, Marius Korsnes, 'Ambition and ambiguity: Expectations and imaginaries developing offshore wind in China', pages no. 50–58, Copyright (2016), with permission from Elsevier.

References

4coffshore, 2013. 4C Offshore Wind Farms Database. Available at: www.4coffshore.com/windfarms/windfarms.aspx?windfarmid=CN46 (Accessed 2 November 2013).

Beaumont, H., 2019. Doing business in China's offshore wind market. Azure International. Presented at GWEC Webinar, 20 May 2019.

Borup, M., Brown, N., Konrad, K. and Van Lente, H., 2006. The sociology of expectations in science and technology. *Technology analysis & strategic management*, 18(3–4), pp. 285–298.

Brown, K. and Bērziņa-Čerenkova, U.A., 2018. Ideology in the era of Xi Jinping. *Journal of Chinese Political Science*, pp. 1–17.

Carlsson, B., 2015. The world according to Xi Jinping. Available at: www.theatlantic.com/international/archive/2015/09/xi-jinping-china-book-chinese-dream/406387/ (Accessed 6 January 2016).

CREO, 2018. China renewable energy outlook 2018. Report prepared by the Energy Research Institute of Academy of Macroeconomic Research/NDRC and the China National Renewable Energy Centre.

CWEA, 2018. Chinese Wind Energy Association (CWEA), Report. Available at: https://community.ieawind.org/HigherLogic/System/DownloadDocumentFile.ashx?DocumentFileKey=46f86d9b-7b90-ab73-f974-e23f97039fb1 (Accessed 1 May 2019).

Deign, J., 2019. China set to overtake UK as offshore wind leader by 2021. Available at: www.greentechmedia.com/articles/read/uk-set-to-lose-offshore-wind-crown-to-china-by-2021 (Accessed 10 May 2019).

Dodson, B., 2012. *China Fast Forward: The Technologies, Green Industries And Innovations Driving The Mainland's Future*. Singapore: John Wiley & Sons.

Economy, E., 2010. *The River Runs Black: The Environmental Challenge to China's Future*, 2nd ed. Ithaca: Cornell University Press.

Geall, S. and Ely, A., 2018. Narratives and pathways towards an ecological civilization in contemporary China. *The China Quarterly*, 236, pp. 1175–1196.

Gosens, J. and Lu, Y., 2013. From lagging to leading? Technological innovation systems in emerging economies and the case of Chinese wind power, *Energy Policy*, 60, 234–250.

Gosens, J. and Lu, Y., 2014. Prospects for global market expansion of China's wind turbine manufacturing industry. *Energy Policy*, 67, 301–318.

Gosens, J., Kåberger, T. and Wang, Y., 2017. China's next renewable energy revolution: goals and mechanisms in the 13th Five Year Plan for energy. *Energy Science & Engineering*, 5(3), pp. 141–155.

GWEC, 2018a. Latest update on China offshore wind. Global Wind Energy Council (GWEC). Available at: https://gwec.net/latest-update-on-china-offshore-wind/ (Accessed 10 May 2019).

GWEC, 2018b. Annual market update 2017. Global wind report. Global Wind Energy Council (GWEC).

GWEC, 2019. Global wind report 2018. Global Wind Energy Council (GWEC). Available at https://gwec.net/global-wind-report-2018/.

He, Z.X., Xu, S.C., Shen, W.X., Zhang, H., Long, R.Y., Yang, H. and Chen, H., 2016. Review of factors affecting China's offshore wind power industry. *Renewable and Sustainable Energy Reviews*, 56, pp. 1372–1386.

Heilmann, S., 2010. Chapter 6: economic governance: authoritarian upgrading and innovative potential. In J. Fewsmith (ed.) *China Today, China Tomorrow: Domestic Politics, Economy, and Society*. Lanham, MD: Rowman & Littlefield, pp. 109–129.

IRENA, 2019. Statistics time series. Trends in renewable energy. By the International Renewable Energy Agency (IRENA). Available at: https://irena.org/Statistics/View-Data-by-Topic/Capacity-and-Generation/Statistics-Time-Series (Accessed 22 May 2019).

Jasanoff, S., and Kim, S.H., 2009. Containing the atom: Sociotechnical imaginaries and nuclear power in the United States and South Korea. *Minerva*, 47(2), pp. 119–146.

Kong, B., 2009. China's Energy Decision-Making: becoming more like the United States? *Journal of Contemporary China*, 18(62), pp. 789–812.

Korsnes, M., 2014. Fragmentation, centralisation and policy learning: An example from China's wind industry, *Journal of Current Chinese Affairs*, 43, pp. 175–205.

Korsnes, M., 2016. A sustainable Chinese catch-up? Product quality and interactive learning in the offshore wind industry. *International Journal of Technological Learning, Innovation and Development*, 8(2), pp. 172–196.

Lewis, J., 2004. Learning how to ride the wind: The disappointments and potential of wind power in China. *China Environment Series*, 6, pp. 84–88.

Lewis, J., 2013. *Green Innovation in China—China's Wind Power Industry and the Global Transition to a Low-Carbon Economy*. New York: Columbia University Press.

Li, J. and Wang, X., 2012. Energy and climate policy in China's twelfth five-year plan: A paradigm shift. *Energy Policy*, *41*, pp. 519–528.

Li, J., Gao, H, Shi, P., Shi, J., Ma, L., Qin, H. and Song, Y., 2007. *China wind power report 2007*. Beijing: China Environmental Science Press.

Li, J., Cai, F., Qiao, L., Xie, H., Gao, H., Yang, X., Tang, W., Wang, W. and Li, X., 2012. China Wind Energy Outlook 2012. Chinese Renewable Energy Industry Association, Greenpeace, Global Wind Energy Council, Chinese Wind Energy Association, Beijing.

Naughton, B. 2007. *The Chinese Economy. Transitions and Growth*. Cambridge, MA: MIT Press.

Naughton, B. Chapter 4: economic growth: from high-speed to high-quality. In J. Fewsmith (ed.) *China Today, China Tomorrow: Domestic Politics, Economy, and Society*. Lanham, MD: Rowman & Littlefield, 2010.

Quartz+Co, 2013. Offshore Wind China, Report. Available at: www.mecintelligence.com/uploads/Offshore wind China-Market Insights.pdf., Quartz+Co. (Accessed 25 November 2013).

Saich, T., 2011. *Governance and Politics of China* (3rd ed.). Houndmills, Basingstoke, Hampshire/New York: Palgrave Macmillan.

Shi, W., Markóczy, L. and Stan, C.V., 2014. The continuing importance of political ties in China. *Academy of Management Perspectives*, 28(1), pp. 57–75.

Smith, P., 2014. China slashes offshore projections. Available at: www.windpoweroffshore.com/article/1318542/china-slashes-offshore-projections?HAYILC=TOPIC (Accessed 14 May 2015).

Stensdal, I., 2014. Chinese climate-change policy 1988–2011: Moving on up. *Asian Perspect*, 38(1), pp. 111–135.

WWF, 2010. *China: An Emerging Offshore Wind Development Hotspot*. Beijing, Oslo: Prepared by the Chinese Wind Energy Association (CWEA) and the Sun Yatsen University. Available at: http://awsassets.wwf.no/downloads/china___an_emerging_offshore_wind_hotspot_cweasyuwwf_aug2010.pdf.

Yang, J., 2014. China unlikely to hit offshore target. Available at: www.windpoweroffshore.com/article/1286062/china-unlikely-hit-offshore-target (Accessed 14 May 2015).

Ydersbond, I. and Korsnes, M., 2016. What drives investment in wind energy? A comparative study of China and the European Union. *Energy, Research & Social Science*, 12, pp. 50–61.

Yu, G., 2019. Doing business in China's offshore wind market. Director of Industry Research, Chinese Wind Energy Association (CWEA). Presented at GWEC Webinar 20 May 2019.

Zhang, D., Zhang, X., He, J. and Chai, Q., 2011. Offshore wind energy development in China: Current status and future perspective. *Renewable and Sustainable Energy Reviews*, 15(9), pp. 4673–4684.

7 China's urban and rural distributed PV development

As we have seen in previous chapters, China's investments in solar PV have been explosive. By the end of 2018, total nationwide PV capacity reached 174 gigawatts (GW), of which 123.8 GW was centralised and 50.6 GW, or about 29 percent, was distributed PV (chinaenergyportal.org 2019). For sure, 50.6 GW might not mean much to you. Knowing that each panel produced on average 1,115 full-load hours of electricity at the end on 2018, this means that 50.6 GW generated about 56 terawatt-hours (TWh) equivalent to the annual power consumption of all of Greece. If we calculate these panels into the area they cover, it equals to about 717,000 football fields, or roughly the size of Trinidad and Tobago. Needless to say, this entails a large change, since someone must have welcomed and integrated all these new solar panels into their backyards and rooftops. Since 2016, distributed solar PV has been growing faster than centralised PV in China (IEA 2017) (see Figure 7.1). Simultaneously with the development of distributed PV, China has also experienced an explosion of cities with 'smart' or 'eco' ambitions since 2011 (Williams 2017). These two developments appear to fit nicely together, but as this chapter shows, these developments are not connected.

The development of distributed renewable energy in China is associated with several important issues in Chinese policymaking and development. Some of these are: ensuring a transition to a low-carbon society, stable electricity provision for its citizens, local job and value creation, challenges with increased inequality, climate change and power market reform. Issues such as these contribute to make clear why China is interested in increased renewable energy, but also why the implementation of such projects is not necessarily straightforward. China is not alone in attempting to solve the challenges which moving towards more decentralised renewable energy generation entail. Only a few areas globally, such as Germany and California, have a density of distributed solar PV comparable to China. Therefore, what is happening in this area is still rather new and untried, and China has the opportunity to pave the way for others to learn.

As pointed out by Gao and Rai (2019), local renewable energy projects should be of great interest to governments—given that local knowledge

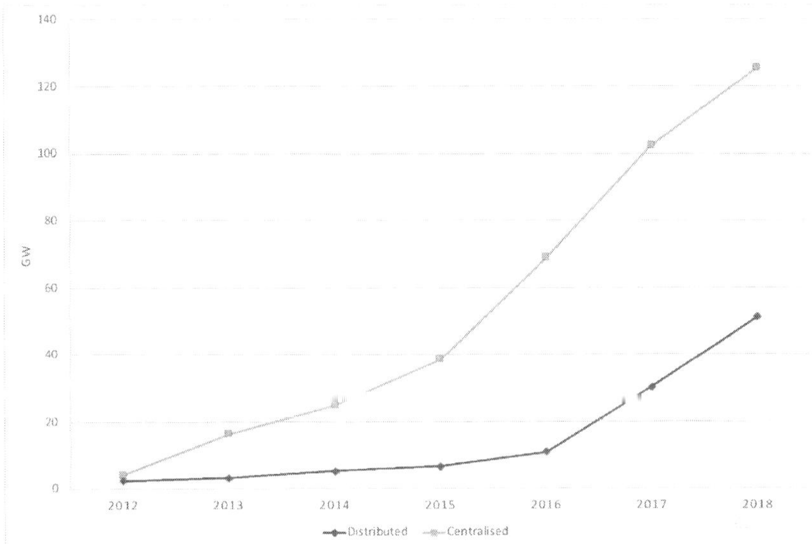

Figure 7.1 Solar PV total distributed and centralised installed capacity 2012–18, in GW.

Source: Data gathered from chinaenergyportal.org (2019).

about, for instance, climate patterns, weather conditions, local organisational structures and avenues for securing financing gives local firms an advantage which may result in local value creation. The role of PV panels in society is not fulfilled when they are rolled out of the factory doors. To actually make use of these panels there is a large chain of activities (typically referred to as balance of system, or BOS), such as wiring, installation, monitoring, grid-connection, maintenance, cleaning and similar. These needs are often serviced by local companies and make up an important aspect of solar PV innovation and value creation that are more relevant in local and small-scale projects (Gao and Rai 2019). An increasing share of PV jobs is related to this segment. In 2017, about 800,000 people were working with PV panel installation and associated tasks in China (IRENA 2018). By presenting the development of distributed PV and contrasting it to China's 'smart' and 'eco' city initiatives, this chapter delves into the gist of these issues and attempts to untangle some relevant facets to help understand better the role of distributed solar energy in China.

Distributed solar PV projects can be defined in a variety of ways. The Chinese Renewable Energy Outlook 2017 pointed out three aspects which distinguish distributed PV from centralised PV. One, the project is small with installation sizing between the kilowatt level and the megawatt level. Two, the project is close to the electricity demand, and the power generated is mainly for local consumption accessed at the power distribution network (below 35kV). Three, the project includes technologies for both sides of supply and demand, such as smart metres and similar (CREO 2017, p. 557).

The official Chinese definition provided by the National Development and Reform Commission (NDRC) was that distributed PV projects are located at or close to the consumption site, where the electricity is self-consumed and the surplus is sold to the power grid, and with sizes smaller than six megawatts (MW) (Zhang 2016). In a 2014 memo titled 'Further implementation of existing distributed PV policies', this size restriction was increased to 20 MW, and the memo included guidelines clarifying issues of responsibilities regarding grid connection as well as identifying rooftops as a preferred mode of installation (Zhang *et al.* 2015, NEA 2014). In line with this, our concern in this chapter is with PV panels which conform to these criteria, that is, small projects which can be installed on rooftops or the ground, producing electricity which is either used onsite or sent to the local electricity grid. This excludes projects which are centralised, here defined as projects above 20 MW of size, and where the electricity primarily is not consumed locally.

In terms of geographical spread, most of the centralised solar PV installations in China are in the northwest regions although most of the electricity is consumed in the East (see Figure 7.2). Due to the increase in distributed PV, generation is now shifting to the Central and Eastern regions, and the provinces which are increasing the fastest are Zhejiang, Shandong, Jiangsu, Anhui and Jiangxi (IEA 2017, p. 64). This change means that the electricity does not have to be transported over long distances. Keep in mind that most of China's energy (including both heat and power) is consumed by the industry sector (61 percent). In fact, at the end of 2016, only 14 percent of the energy was consumed in buildings, and 21 percent related to transport (CREO 2017, p. 11). Moreover, coal was used for 67 percent of the electricity generation (CREO 2017), implying that the potential of distributed PV for emissions reduction is larger for the industry sector. Also, since electricity prices are higher for industry and commercial sectors than for residential sectors, it is more interesting for industry and commercial sectors to reduce these costs.

For distributed PV to take a significant role in changing electricity consumption, the power grid needs to be made flexible and adjusted for smaller-scale production. This is currently a large challenge in China's electricity provision:

> (…) with rapid increase of distributed renewable energy in recent years, it's hard to fully utilize distributed resources to achieve the optimal technical and economic effect with respect to active distribution networks. Additionally, traditional distribution network lacks technological means for active regulation and control, which makes distributed power sources hardly participate in system voltage and reactive control, and thus hardly provide auxiliary services.
>
> (CREO 2017, p. 487)

As this quote says, the rigidity of the distribution network makes it hard to make use of distributed power production for balancing and optimising the

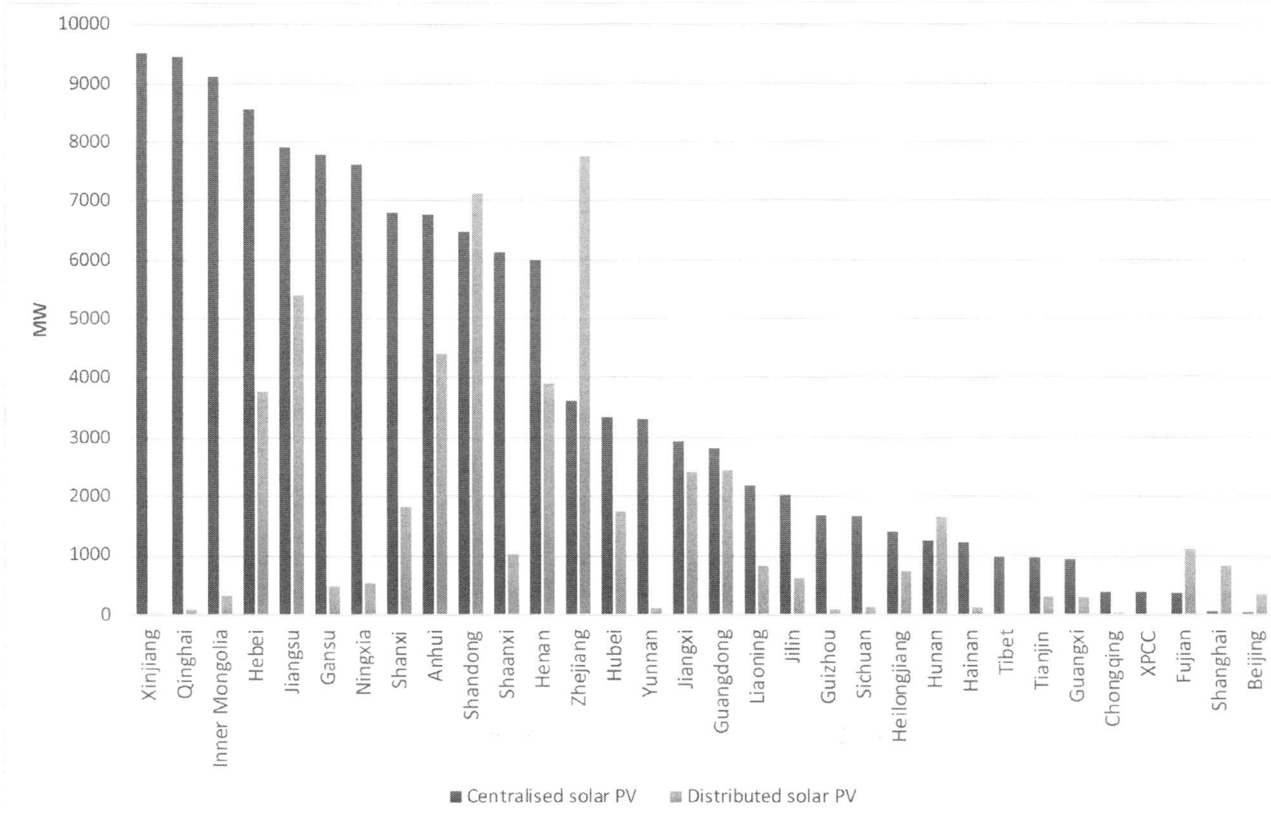

Figure 7.2 Cumulative centralised and distributed solar PV in China, end of 2018, in MW, by province.

Source: chinaenergyportal.org (2019).

grid, thus making them unable to fill their role in supporting grid services. This illustrates that the increasing use of distributed PV presents larger-scale systemic issues related to policy and governance of supply and demand. The aim is to go into more detail into this kind of issue, investigating the reasons for increased use of urban and rural distributed PV in China, as well as consequences this development has for further operating local grid networks, changes in energy use and continued decarbonisation. In this context, the chapter will provide answers to these questions: what are the reasons and processes behind the increased use of distributed PV in China? What are the impacts of this decentralisation of electricity production on the operation of electricity grids? To what extent is this development connected to the explosion of 'smart' and/or 'eco' cities in China?

Distributed renewable energy and 'eco' cities in China

Decentralised electricity

In general, the increasing use of renewable energy locally has several impacts relating to electricity dispatch and grid, electricity pricing, energy governance and policy as well as electricity consumers and energy use (Davidson 2018). Much research has already been conducted on the role of centralised wind and solar energy in general and the various effects they may have on decarbonising China's electricity supply (see, e.g., Qi, Zhang and Karplus 2014; Kirkegaard 2019; Davidson *et al.* 2016). However, the decentralisation of electricity production, which an increase of distributed solar PV implies, can potentially have wide-ranging and radical influences we as of yet do not know the extent of. Recently, some focus has been directed at the role of market reforms in China, and the role which spot prices on electricity might play to reduce the amount of renewable electricity which is wasted (e.g., Zhang, Andrews-Speed and Li 2018). Such studies give indications as to why decentralised solar PV provision solutions face challenges, all whilst selling electricity back to the grid remains difficult or impossible in most of China. With today's system (before 2020), with some exceptions for large industrial consumers, there is no possibility to trade electricity between consumers and producers. We will return to this later in this chapter and the next chapter. Until now, the main emphasis of research has been on large-scale, centralised installation and usage of renewable energy. Large-scale renewables arguably has different implications for energy systems and energy use than small-scale PV since, even though they pose issues of intermittency as well, they resemble traditional centralised electricity provision systems to a larger extent.

In a non-Chinese context, some research has been done on the energy system and flexibility effects (e.g., Koch, Girard and McKoen 2012; Kubli, Loock and Wüstenhagen 2018), energy use impacts (e.g., Ellsworth-Krebs and Reid 2016) and low-carbon consequences (e.g., Korsnes, Berker and Woods 2018) of increased use of distributed renewable energy systems, such as rooftop PV. Such studies emphasise the role 'energy prosumers', that is, local producers and consumers of energy (both heat and electricity). Energy prosuming could entail a change in how electricity has been generated traditionally, a potential challenge to established power utilities and how energy is used (Ellsworth-Krebs and Reid 2016). In a Chinese context the focus on smaller-scale micro-production of renewable energy has been confined mainly to off-grid, agriculture and electrification projects in less developed areas which lack access to electricity (e.g., Wallace, Jingming and Shangbin 1998; Xue 2017; Geall, Shen and Gong 2018). Projects on rural electrification were important in the very beginning of China's solar PV experience in the 1990s, and projects such as the 'Brightness programme' (launched in 1998), the Township Electrification Programme (launched 2001), and the Village Electrification Programme (launched 2006) were early initiatives focusing on off-grid areas (Marigo 2007). In 2003, such rural projects contributed 28 MW, which at the time was more than half of the total installed capacity in China (Marigo 2007). In 2013, 99.8 percent of China's population had access to electricity, up from less than 10 percent in 1949 (Geall, Shen and Gong 2018). As Xi Jinping came to power in 2012, poverty alleviation became elevated to one of the highest development priorities for China (Geall, Shen and Gong 2018). In 2014 the government launched the solar energy for poverty alleviation ('Poverty Alleviation'), which until 2020 aims to add more than ten GW of distributed solar PV projects benefitting more than two million households across China (Geall, Shen and Gong 2018).

Some recent work has studied distributed solar PV policy and development in China. Zhang (2016) described the more formal policy mechanisms, as well as the challenges which distributed PV projects have faced in China. Such concerns were in 2015 amongst others related to: 'building (and/or rooftop) ownership problem, the unattractiveness of on-grid tariff and the low proportion of self-consumption, the barriers to grid connection, and the difficulty in obtaining financing' (Zhang 2016, p. 99). Yuan *et al.* (2014) and Luo *et al.* (2016) looked closer at the various financing risks involved in distributed PV projects, and also found similar issues: definition of rooftop property rights and access, gaining access to the grid, and these issues lead to uncertainties connected to the profitability of the projects. Zhang *et al.* (2015) confirm some of these uncertainties but points out another issue related to the self-consumption FiT model which was used. They found that if all the electricity generated had been self-consumed the PV projects would have a higher profit than if all electricity was sold to the grid. However, as it was uncertain if companies would consume the electricity themselves or not, financial agencies would not invest in developing the rooftop PV system.

Several of the above-mentioned challenges still exist today. However, given the explosive growth in distributed PV the past years, several of them may have been overcome. Indeed, as pointed out in a recent study by Gao and Rai (2019), provincial governments have prepared the ground for increased distributed solar PV systems since 2011 and have also developed agencies and procedures which assist in solving problems such as rooftop ownership and similar. For example, the Jiangsu government 'assesses all available and suitable rooftop resources for distributed PV and helps address the problem of unclear rooftop ownership, (…) and requests Grid Company to provide specific services to help costumers connect distributed PV electricity to the grid' (Gao and Rai 2019, p. 366). What is more, not much research has been conducted on a more urban and developed context of distributed PV projects in China. Wu, Zuidema and Gugerell (2018), for instance, conducted a recent case study of four Chinese cities, and found that local governments had only modest willingness and ability to adapt to local, decentralised renewable energy solutions. Hence, given the divergence between actual installations of distributed PV and the modest willingness to adapt, it is still a puzzle why the development has come about.

The increasing use of distributed PV systems is a new phenomenon globally, and for the Chinese case the starting point, as will be the case here, must be to describe the conditions within which it develops as well as the local challenges and benefits associated with the increasing presence of distributed PV systems. In this respect, a comparative analysis which looks at motivations behind rural and urban distributed PV projects in China can give insights into the reasons and processes behind the increased use of distributed PV in China.

'Smart' and 'eco' cities in China

The term 'eco-city' is a quite loosely defined concept with no universal agreement as to what it actually contains (Williams 2017). In China it seems like there has been an inflation of a variety of initiatives using labels such as 'eco', 'green', 'low carbon' and 'smart'. Many of these labels are used interchangeably, making it difficult to specify what aspect of 'smart' or 'eco' is being studied in each case. Williams (2017) points out that although the number of eco-cities in China have exploded the past years, we still no not know exactly what that entails with respect to sustainability:

> Many new Chinese Eco-cities aren't really new but are existing cities with add-ons. Some Chinese Eco-cities are not really cities. Many Chinese Eco-cities aren't even 'eco' in any acceptable Western sense of the word. Very often a Chinese Eco-city is a park, or a well-insulated industrial zone, or a waste incinerator—whatever it takes to tick the sustainability consultant's box, get the certificate, and rub eco-critics' noses in it.
>
> (Williams 2017, Kindle loc. 3767–3770)

The notion of 'smart' city has also exploded the past years. In November 2012, the Ministry of Housing and Urban-Rural Development (MOHURD) issued a 'Notice on carrying out national pilot smart cities' where city governments were urged to form plans on pilot smart cities with a focus on security systems, smart construction, administration, industry and economy

(Tan-Mullins *et al.* 2017). After that, several new smart pilot cities were launched consecutively: In January 2013, 90 national pilot smart cities launched; in April 2014, 103 more were launched; and in April 2015, 277 pilot smart cities were launched. There have been several studies on the role and development of both smart and eco-cities in China. De Jong *et al.* (2016) addressed the policy implementation gap in Chinese eco cities and found that a too strong reliance on top-down technocratic planning made the local sustainability efforts less grounded and local governments less engaged. Therefore, they suggested that paying closer attention to the interdependency between the interests of local government actors and the interests of land and real-estate developers might assist the implementation of real eco cities in China. Other studies which have taken a more bottom-up approach to study Chinese eco cities [illegible] that issues such as inequality and the 'new urban poor' also must be taken into consideration to make Chinese eco-cities successful (Caprotti 2014). Studies along this line of thinking have also emphasised the importance of focussing on the actual lived experiences of residents in these various urban sustainability experiments focussing more on human needs and comfortable cities, and not only on plans, visions and blueprints (Caprotti and Gong 2017; Caprotti 2018). In sum, however, it appears that 'eco' projects are not very distinguishable from regular urban development projects in China.

Cheshmehzangi *et al.* (2018, p. 65) look closer at 'low carbon' cities and point out that in China, the term was coined to 'blend the low carbon planning and the eco-city initiatives'. Moreover, they note that in recent years Chinese low carbon cities have given more attention to developing key performance indicators (KPIs) for actual decarbonisation with a focus on big data relating to energy demand and various sources of CO_2 production. This would then potentially move 'low carbon' projects closer to the realm of 'smart city' projects, which are more related to digitalisation and usage of information and communication technologies (ICTs). Given that solar PV is a low carbon technology, the focus on low carbon indicators would make space, in theory, for an integration of renewable energy. As we will see soon, thus far it appears that this has not been central to urban decarbonisation efforts in China. Although there is no unified definition of what smart grids and systems entail, in a European context we can generally say that it implies an increased integration of new ICTs which enable communication between different actors and integrating new actors and assigning new roles to existing actors in the electricity system, for instance, where households both produce and consume electricity (Christensen *et al.* 2016, p. 6). In other words, if Europe is to be taken as a benchmark, studying China's smart cities could include looking for technologies which are not representing business as usual, such as smart grids, smart metres or big data analysis tools. Moreover, when electricity increasingly is produced locally, it becomes interesting to see whether implicated stakeholders assume new roles in the electricity system. Thus, we need to develop a better understanding the roles of users and other stakeholders of Chinese distributed PV projects.

Rural and urban distributed PV: a small taste

This section first describes one rural and two urban cases of decentralised PV projects emphasising how they are received and what types of needs they address for their users. We then take a step back and discuss the various reasons why Chinese PV installations have moved from centralised utility-scale to smaller-scale distributed installations, the connection to smart and eco-cities, and some of the more general challenges which have been faced.

The rural distributed PV systems

Shanghe is a county with around 560,000 people in Shandong Province. To investigate the use of rural distributed PV, I went, together with four students from Tsinghua University, to a town called Huairen, with around 30,000 people, divided into 58 smaller villages. The first day we were there, we met the local county mayor (of Shanghe) and the local economy representative of the county, as well as the local government officials of Huairen, and they presented current status with respect to renewable energy, as well as their plans for the future. During the first day we also visited the local grid and dispatch centres, several smaller ground-mounted PV systems, a centre which monitored all the solar PV production, a 40-MW solar PV farm, and a district-heating centre providing heat for the village. The next day we visited ten households, of which six had rooftop PV, and all had solar hot water. The Tsinghua students were asking questions related to heating and general energy and space use in the households such as for cooking, washing and so on, and I added questions particularly about the rooftop PV and solar hot water systems.

Upon arrival in Huairen village, we met with the local government who gave a presentation about their renewable energy strategies. It was clear that they had grand visions for the future, with ambitions to make Huairen a leading green town in China, which would attract national recognition. The main reason they were interested in this was that the central government had made renewable energy more important and they wanted to be a model town and get attention nationally. They wanted to be an example to follow and become more 'advanced' and 'greener'. Also, it became clear that the local mayor wanted to use this as an opportunity to make a career. What they had achieved so far was some rooftop and ground-mounted PV, the 40-MW large-scale PV station, geothermal heating system for the winter providing heating for those living in the centre, 50 MW of wind power, and smart systems where people could monitor their own production. For the future, this would only increase, and their aim was to combine several different renewable energy sources together and, ultimately, to become self-sufficient. Another aim was to bring in more industry to help with local development.

When asked what the local population thinks of the increasing use of renewable energy locally, they answered that the local population was very happy with the rooftop PV solutions. It was very simple for the households:

one-stop shop companies rented their roofs and paid them money. Each house typically installed 5.4 kWp per roof. The household had to pay 29,000 RMB (about 3,700 euro) for the total installation, and they would earn between 6,000 to 7,000 RMB in income per year so in about five years the investment has been paid down. For this investment households invested their own money and did not have to get a loan. According to the Huairen mayor, people were willing to use their own money because 'this town is rich compared to other towns, and people here have a high regard of PV'. Although the subsidy was recently reduced from 0.48 to 0.37 RMB per kWh the projects still paid off. I also asked if the increase in use of PV was leading to more local jobs, but the 40-MW project and other distributed PV projects locally had only led to 10 to 20 new jobs for cleaning the PV panels and similar, so not many extra jobs.

People in Huairen village appeared to use quite a lot of their income on the house. The ten different households which we visited were relatively similar with respect to their installations and economic schemes. A third-party company took care of everything from installing the solar panels to dealing with grid connection. All the electricity was sold to the grid, and the grid company made payments to the households four times per year. When asked if there was any change in their consumption, they answered that there was no change at all, they didn't really think much about it. They could follow the power production on an app, but they never really used it. In fact, none of the households we visited were able to get the app to work. Thus, the households did not see the role of such smart, mobile phone-based overview-mechanisms as very interesting. Nevertheless, they did sometimes have a look at the inverter itself to see the production, but this too was not really something they cared much about.

The production of electricity was usually around six hours per day. Since the PV panels had a life expectancy of 25 years and the investment would be paid down after around 5 years, they could enjoy an income the next 20 years. This extra income was the main motivation for all the visited households for installing these rooftop PV systems. One of the households, occupied by a single elderly woman, had just one kWp installed, and she had gotten it for free as part of the national 'Poverty Alleviation Programme'. She got around 2,000 RMB every year. In general, she didn't seem to know so much about how it worked or what the panels did, she was just happy that she got this small extra income.

The urban distributed PV systems

Moving to the two urban sites visited in Beijing, it became apparent that different types of matters were of concern compared to the rural cases. Whereas the rural cases all sold electricity directly to the grid and did not concern themselves so much about the production apart from the revenue which it generated every third month, the larger, urban installations faced more hurdles. The two cases were an indoor ski resort (location 1) and a design and research institute (location 2), both located at the outskirts of Beijing. Location 1 had a 300-kWp installation, owned by a PV company, and location 2 had 500 kWp, owned by the design and research institute. Both of the projects were initiated in 2015 because the project owners wanted to get a head start with the experience of installing and managing rooftop PV projects, which could give them a market edge in the future. None of them experienced any difficulties with connecting their installation to the grid. None of the two had noticed any difference in the energy use of the building after the rooftop PV was installed, apart from the fact that the energy use during the lunch-peak was reduced somewhat.

For financial government support, the companies could 'pick one of two' mechanisms: to receive a feed-in tariff and sell all electricity to the grid (no self-consumption), or to use all electricity and get a subsidy, but to sell the remaining surplus electricity to a lower price (Zhang 2016). The project engineer of location 2 (Interview #74, Jun. 2018), who had picked the option to use all electricity, said that:

> If we do not use all our electricity, for instance under the Spring Festival or similar, we sell it to the grid, but to a lower price. It is only [0.35] RMB, so it is not attractive for us to sell it there, since our electricity price is [1.26] RMB. We would prefer to use it ourselves. Now we sell around only one percent of all the produced electricity to the grid.

This means that the company had to make relatively detailed calculations about their use versus non-use, and what they would earn or lose if they picked one of the options. Given the high electricity price, they could save quite some money if they used the electricity themselves.

In terms of other concerns these projects raised, the issue of fire safety on the rooftop was prominent and a source of some concern for the local project engineer at location 2 (Interview #74, Jun. 2018):

> If it catches fire, then we are not allowed to put it out. This is because of the risk with the electricity production. So, if it catches fire, we have to turn off the power, and just wait. But we do several things to make sure that there will be no fire. We don't keep anything on the roof that can catch fire, small things or trash, and we keep the panels clean and make sure there are no dust particles or dry leaves or similar. We also have an insurance on the installation, which we got after the installation was finished.

This situation thus appeared to generate a somewhat grim situation for the company, should a fire occur. Fire safety therefore became more important for this company than before the rooftop PV had been installed. Another issue which was important was related to keeping the solar PV panels clean so that they could produce optimally. In location 1, the project manager explained that they had quite complex algorithms to decide when it was the best time to clean the panels, as it depended a lot on the weather conditions. A general principle was to clean them once per month, but to save money it could, for instance, be decided to postpone cleaning if heavy rain (which would clean the panels) was expected within the next week. They therefore had invested in a quite comprehensive weather station and monitoring system, which would allow them to know when to send out a project engineer to the site.

Opportunities and drawbacks with the increase of distributed PV in China

We have now seen some particular cases of how urban and rural distributed PV experiences have been made in China. To be sure, these cases are not generalisable to all of China, but they do provide an understanding of the way in which some distributed PV projects in China have come about the past years. This section goes on to discuss more in detail the reasons for the increase of distributed solar PV projects, what types of challenges they have faced, and the lacking affiliation these projects have had with China's 'smart' and 'eco' city developments. As we will see shortly, it turns out that this issue is strongly connected with the larger problems of establishing a smart electricity grid.

From the very beginning of the data collection period in 2018, I started out with the assumption that the right place to seek distributed renewable energy projects would be at the various 'smart' and 'eco' cities which exist around China. This turned out to be quite mistaken. Every person I talked with was unable to mention a 'smart' or 'eco' city project that also had distributed PV as an integrated part, even only a small part. Naturally, there were some examples, such as the Tianjin Eco-City project, but the amounts of distributed PV there were minuscule compared to the decarbonisation efforts it was attempting to achieve. One Chinese professor (Interview #76, Aug. 2018) explained the reason for the lacking connection in this way:

> Because the government departments are divided, they don't have much collaboration: some work on gardens, some on roads, some on renewables. This will be some learning for the future, and this is something we can improve on. We understand that the different departments are interrelated, but in the actual world it is very difficult to integrate them.

Another interviewee from a government urban planning bureau (Interview #59, Apr. 2018) pointed out the focus on economic growth as one of the main reasons for not including distributed PV: 'Several of these smart/eco projects are more focusing on economy and economic growth. These projects seem fantastic, but when you look into it, there's not really much new happening'. As expected, therefore, the fragmented nature of the Chinese government as well as a singular focus on quests for economic growth were identified as reasons for the lacking connection between the increase in distributed PV projects and eco and smart city projects.

Nevertheless, this does not explain why distributed PV still grew so quickly after 2015. As it turned out, several reasons were indicated. For one, distributed solar PV installations became more popular as a response to a decline in the profitability of centralised PV projects. As pointed out by one energy advisor (Interview #63, May 2018):

> The high degree of curtailment (in utility scale PV projects) led to lower profits than expected, meaning lower incomes. This led to that several companies could not cover their costs: supply chain companies, such as wafer producers and so on. This was of course connected to the fact that these large-scale PV projects required a large investment, and it became more difficult to get big loans for such big projects. So more small-scale, distributed projects became more interesting, because they were easier to finance. Also, most of the cheapest land had been taken, so land-investment became a larger part of the cost of these projects.

Another important role in increasing the interest in distributed PV projects was the government decision to reaffirm the intention to reform the electricity sector after the 2015 'Document no. 9'. This document promised to improve the prospects for energy trading, making distributed installations more interesting, e.g. by making it possible to sell electricity to neighbouring companies. A government official pointed this out (Interview #69, May 2018):

> We know that a lot of rooftop PV systems of factories, in the weekend they sometimes cannot use all the production, so they want to promote a more open market, open trade, and it is also a sort of energy production and consumption revolution. I think distributed solar PV and wind can be a good approach to promote this energy production and consumption revolution. And another thing is, in the past, if we wanted to sell the surplus electricity to the city, they could only get the coal generation feed-in tariff locally, and that is not very high, so maybe they can sell to their neighbours. Maybe the neighbour is willing to pay a little higher than that, as long as it is lower than the grid price.

Hence, selling electricity to neighbouring facilities would be a way to solve the low profit that would be gained from selling electricity to the grid. Nevertheless, reform of the electricity sector is a slow process, and the lack of a spot-price market of electricity made it difficult to find prices which would reflect a balance between costs and benefits.

New additions of distributed PV were growing too quickly, and on 31 May, 2018 (referred to by industry stakeholders as '31.5') the Chinese government decided to halt the support of centralised PV projects (except from the so called 'frontrunner' or 'leader' bases) and to put a ten-GW limit on total distributed PV installations. Poverty alleviation projects were not affected by the policy change. The 31.5 incident appeared to be expected by energy analysts, although industry actors were taken by surprise by the sudden change. Several events had been leading up to the change, such as uncontrolled growth, large delays in paying out promised subsidies, and fear of over-funding the PV industry, but many experts also pointed to the role of State Grid. Indeed, the lacking link between the distributed PV development and 'smart' and 'eco' cities appeared to be exist also because of the reluctant role which State Grid played in providing smart solutions for distributed PV systems. One of the issues were relating to connecting PV systems to the electricity grid. An experienced project manager from a PV installation company pointed this out (Interview #76, Jun. 2018):

> In principle it is a rather simple thing to get connection to the grid, especially in small residential areas. They (State Grid) are mandated to connect all renewable energy, so they have to do it. And it is not a complicated thing. But of course, State Grid is not so interested in PV because coal power is better for them and pays off more. PV is more expensive, also because of the lag in payments from the central government. What the grid company really cares about is two main things: First, the quality of the electricity delivered, and here stability actually weighs in somewhat, and second, the capacity of the sub-station, which sometimes can be a barrier for distributed projects.

In sum, coal power was more stable, and thus easier to deal with, and because distributed solar PV projects implied production inside load centres the local sub-stations often had to be upgraded to be able to handle the extra strain.

Although connecting solar PV systems to the electricity grid was a simple matter in principle, it could take a long time for them to be connected and ready to sell electricity back to the grid. This was especially a problem for Beijing. As one energy specialist explained: 'the grid company can take a long time to arrange the system for sales back to grid, between 45–150 days. Beijing is generally too conservative here, and it is mainly the grid company that is resisting' (Interview #66, May 2018). In an interview with a central government official (Interview #32, Dec. 2013), it was clear that reform in general was necessary to make things work. Moreover, the hidden costs of electricity transmission, something that only the grid company really knows, emerged as one particular issue:

> It is not clear what the transmission fee structure is. For coal fired power plants, that is very clear. How much coal they use, what is the coal fire plant technology, it's very clear. So normally, at this price you can make a benefit, and at this price you have a problem, that is clear. But for the transmission lines, this is a problem, it is not clear. (...) The end user price includes the producer price, transmission fees, and the distribution fees. The producer price and cost is very clear. But the transmission is not clear. So, I pay the electricity bill, for instance in Beijing, for residents, maybe it is [0.5] yuan per kWh. The cost to produce is maybe

> only [0.2], or [0.3]. I don't think the transmission fee is that high, and sometimes it is maybe lower, I don't know.
>
> (Interview #32, Dec. 2013)

Interestingly, even a relatively high-standing government official, whom we could assume would know something about the revenues of State Grid, did not know the actual costs of transmission. Probing further into the reasons behind these hidden costs, I got an explanation from one energy consultant (Interview #67, May 2018): 'Clearly State Grid doesn't want to give up how they set prices, because then people will start asking questions and demand different prices and so on'. This was one of the reasons why a reform of the power system arguably would benefit distributed PV projects in China. Nevertheless, an associated problem was the status of the grid company:

> State Grid is not willing to give up their position easily. T&D [transmission and distribution] prices are negotiated, and it is difficult to come up with a good price. Currently it is mainly attractive with rooftop PV for large office buildings or industry parks and so on—it is beneficial for large customers.
>
> (Interview #66, May 2018)

State Grid therefore faced two interrelated issues: risking losing control of electricity grid transmissions and distribution and thereby losing an important position for negotiation as well as risks of economic losses.

As the brief literature review showed earlier, access to roofs appeared to be a barrier until 2015. This did not appear in my interviews; however, the quality of roofs came up as an issue. One energy consultant pointed out that 'one of the main reasons that distributed PV is difficult is the low quality of rooftops here. Concrete rooftops are good, but not the steel constructions' (NEO 2018). When asked about access to rooftops, the project engineer of location 2 (Interview #74, Jun. 2018), introduced above, pointed out that there was a great potential in offering new types of business deals with rooftop owners:

> The first thing we do before starting a project is to find out who owns the roof, and whether or not we can access it. In general, we use two main approaches: One, we pay a rent to the owners of the roof, or two, we pay part of the earnings from the PV panels to the owner. This is typically for a period of 25 years, and a main part of the agreement is that we must have access to the roof in that whole period. It doesn't really matter who owns the roof, and this can vary from place to place—as long as we have an agreement, a contract, about access. One business model that we have used with some success with hotels, for instance, is that we install the panels on their roof, and we split the incomes into three parts. One for us, one for them, and a third part to their online service-provider, for instance Ctrip, which can be used to reduce the

> prices of living in the hotel. This would attract more customers and might lead to more hotels wanting this type of agreement in the future, so it is a win-win situation.

Here we see that what once was considered a challenge has been transformed into an opportunity. As this project manager pointed out, PV installation companies have been quite clever in coming up with solutions which would create win-win situations for themselves and their customers. In line with Gao and Rai (2019), this is an instance of local innovation which originates from the installation process of PV panels (the BOS), and not the manufacturing process. Such instances of innovation can contribute to a more widespread local benefit of the increase in distributed PV panels compared to the benefits which befall local areas where the panels are produced.

On a more general note, interviewees expressed a concern about who the actual users of these distributed PV systems would be, and who would cover the costs. One energy analyst (Interview #67, May 2018) pointed out that 'there is generally no focus on individual households here'. Another way to delineate the issue was between users who were overly technology interested and had the means, and those who needed the electricity. A professor of low-energy buildings (Interview #70, May 2018) put it this way:

> For rooftop PV, currently, I would say that there are two kinds of users. You have the rich ones, who can afford to install these systems and who can afford a house. You know, in China, most people live in apartments. But rich people might have several apartments in the city, and a big house on the countryside somewhere. They focus on solar PV not for money, but for fun. Also, of course there are some who are highly educated, who might be interested in PV for other reasons, like selling back to the grid for instance or other things. But the main point is that these people do it voluntarily. The second group of users are those who fall under the government's poverty alleviation programmes that aim to help poverty not by handing out money, but by giving people access to electricity through PV.

This statement puts the increased distributed PV development in a new light, distinguishing (perhaps overly) sharply between the very rich, who do it for fun, and the very poor, who do it out of necessity.

As demonstrated in this section, it is not surprising that the connection with 'smart' and 'eco' cities has not been as strong, and that the support schemes for PV have been rolled back. We have seen that the relatively expected problem of fragmentation in governing structures, and quests for economic growth were prominent reasons. Nevertheless, we also identified several challenges with both centralised and distributed solar PV projects, and, as will be discussed in the next chapter, it appears that the government wants to solve these through more market-based mechanisms involving less government support.

Conclusions

This chapter started out by reviewing the current literature on distributed renewable energy and smart cities. Then the chapter presented one rural and two urban cases of distributed PV systems, and the various issues which emerged in these projects. These two cases are examples of what is happening in China's distributed solar PV field, and give a glimpse of a relatively unknown but arguably important aspect of China's energy transition. Lastly, based on expert interviews, the chapter presented some more general arguments relating to what has enabled a surge in distributed PV projects, the challenges these have faced and why they have not been connected to 'smart' and 'eco' city projects. In the specific cases, we observed that there were some very concrete concerns which actual users of rooftop PV installations faced, such as concerns for fire safety, cleaning panels and low usage of smart monitoring technologies, which were not addressed by the various policies and support regimes.

The chapter found that the most important reasons for the lacking connection between 'smart' and 'eco' city projects and the development of distributed renewable energy projects could be divided into three main categories. First, 'smart' and 'eco' city projects have mainly been focussing on economic growth, and issues relating to lacking communication between ministries and government departments hampered connections to renewable energy technologies. Second, the reluctant role of State Grid in reforming the electricity market enabling sales of electricity back to the grid complicated the connection with the 'smart' agenda. And third, the increasing use of distributed PV happened as a response to the lower availability and profitability for *centralised* PV projects, and thus not as a response to a more strategic development of a 'smart' or 'eco' city. This shift in profitability was related to; first, a large amount of electricity waste (curtailment) in large-scale, centralised PV projects, which lowered profits; second, high electricity prices for industry and commercial businesses making the distributed PV investment interesting as a way to save money; third, many of the best solar sites with cheaper land prices for large-scale PV had been taken, making urban, rooftop-locations more attractive. Lastly, we also saw that new business models for hotels and other large companies made it more interesting for them to justify investing in rooftop PV.

In terms of actual decarbonisation efforts in Chinese cities, the relatively simple deduction that can be made from these findings is that the criteria for achieving 'smart' and/or 'eco' cities should also include how electricity is sourced. Also, there should be made efforts to ensure more flexible mechanisms to connect and sell electricity back to the grid, which would increase the amount of distributed PV systems in the cities. Nevertheless, pointing this out may not be enough to make sure that anything will actually happen. As a matter of fact, a number of different types of business models occur in the distributed solar PV field in China, as well as a large amount of local innovation and value creation connected to the installation and operation of distributed PV panels (Gao and Rai 2019). Thus, perhaps the most important insight that the lacking connection between 'smart' and 'eco' city projects and distributed PV projects has shown, is that that bottom-up solutions are just as important for the viability of distributed solar PV projects as top-down initiatives.

This chapter has opened a relatively closed door into the way in which real, local distributed PV projects work in urban and rural areas of China and has shown the motivations behind these initiatives. The urban and rural cases introduced in this chapter indicated that there is a difference in the development in the two settings. Whereas the urban sites were largely focusing on industry and commercial-scale rooftop projects with ambitions to save energy costs, the rural project was initiated by a highly enthusiastic local government, and the local village population were interested in small distributed PV installations to gain an extra income. In other words, the urban projects were more interested in the actual production of power which the PV panels could do, whilst the for the rural project the actual production was relevant only to the extent that it would increase the households' income. Both urban and rural locations had in this respect an interest in maintaining high production. Nevertheless, the problem of air pollution in urban locations such as Beijing and Shanghai reduces the generation output from PV panels. Moreover, the rural case showed a lower interest in smart solutions for monitoring their production. The main concern in the more rural areas had been to get electricity access per se, and the 'problem' of demand response management was not as relevant as it was in the urban case.

In general, as also pointed out by Zhang (2016), the issue of getting the solar PV system connected to the grid still is a problem. This chapter showed that this related amongst other things to the status of the State Grid Company, which inhibited a development towards more smart systems. State Grid was delaying grid connection in several cases and was reluctant to reveal hidden costs of electricity transmission, which will become available once the electricity system is deregulated. It was simply not beneficial for the grid company to restructure, and this will likely take time to change. This means that the 'smart' agenda which is discussed in European countries is still several years into the future in China. Nevertheless, change is happening quickly, as pointed out by one of the government officials which were interviewed (Interview #63, May 2018):

> The smart focus was big in 2011 here, people were talking about smart solutions, smart grids etc. But it was just a decoy: Since China did not have any of these smart solutions, they argued that China could not install much or any renewable energy, but that was entirely wrong. Now, however, it seems that things will move in a more decentralized direction, and then smart solutions might also enter.

Practically all provinces in China are already experimenting with electricity markets (Davidson 2018, p. 64). As shown in this chapter, it appears that

China is moving towards more decentralised power production, and it would not be a great surprise if some areas of China will be leading also in this field in few years. Nevertheless, it is far from certain that these electricity markets will look exactly like the ones we see in Europe. As pointed out by the same government official (Interview #63, May 2018): 'The Chinese government sees markets as tools to achieve certain goals, not as a natural state'. For the CCP, such a goal could be a better distribution of resources as well as reducing the impact of climate change and environmental degradation. The Chinese government obviously does not like that electricity resources are wasted (curtailed) leading to millions of RMB lost. As I will return to in the next chapter, this is an issue which must be solved, and the solution will likely involve a coordination of top-down and bottom-up measures.

References

Caprotti, F., 2014. Critical research on eco-cities? A walk through the Sino-Singapore Tianjin Eco-City, China. *Cities*, 36, pp. 10–17.

Caprotti, F., 2018. Future cities: moving from technical to human needs. *Palgrave Communications*, 4(1), pp. 35.

Caprotti, F., and Gong, Z., 2017. Social sustainability and residents' experiences in a new Chinese eco-city. *Habitat International*, 61, pp. 45–54.

Cheshmehzangi, A., Xie, L. and Tan-Mullins, M., 2018. The role of international actors in low-carbon transitions of Shenzhen's International Low Carbon City in China. *Cities*, 74, 64–74.

chinaenergyportal.org, 2019. 2018 PV installations utility and distributed by province. Available at: https://chinaenergyportal.org/en/2018-pv-installations-utility-and-distributed-by-province/ (Accessed 2 May 2019).

CREO, 2017. China renewable energy outlook 2017. Report by Energy Research Institute of China Academy of Macroeconomic Research/China National Development and Reform Commission, China National Renewable Energy Centre (CNREC).

Christensen, T.H., Friis, F., Ryghaug, M., Skjølsvold, T.M., Throndsen, W., Fernandez, S.R. and Perez, E.S., 2016. Recommendations and criteria for the design of smart grid solutions for households. In *ERA-Net Smart Grids Project Report, Integrating Households in the Smart Grid*. Available at: https://vbn.aau.dk/ws/portalfiles/portal/234919047/WP5_Design_Criteria_report_Final.pdf.

Davidson, M.R., 2018. Creating markets for wind electricity in China: case studies in energy policy and regulation. PhD Thesis in Engineering Systems at the Massachusetts Institute of Technology. Cambridge, MA.

Davidson, M.R., Zhang, D., Xiong, W., Zhang, X. and Karplus, V.J., 2016. Modelling the potential for wind energy integration on China's coal-heavy electricity grid. *Nature Energy*, 1(7), pp. 1–7.

de Jong, M., Yu, C., Joss, S., Wennersten, R., Yu, L., Zhang, X. and Ma, X., 2016. Eco city development in China: addressing the policy implementation challenge. *Journal of Cleaner Production*, 134, pp. 31–41.

Ellsworth-Krebs, K. and Reid, L., 2016. Conceptualising energy prosumption: Exploring energy production, consumption and microgeneration in Scotland, UK. *Environment and Planning A: Economy and Space*, 48(10), pp. 1988–2005.

Gao, X. and Rai, V., 2019. Local demand-pull policy and energy innovation: Evidence from the solar photovoltaic market in China. *Energy Policy*, 128, pp. 364–376.

Geall, S., Shen, W. and Gong, B., 2018. Solar energy for poverty alleviation in China: State ambitions, bureaucratic interests, and local realities. *Energy Research & Social Science*, 41, pp. 238–248.

IEA, 2017. Prospects for distributed energy systems in China. OECD/IEA report, IEA series in support of IEA–China Energy Cooperation. Available at: www.iea.org/publications/freepublications/publication/ProspectsforDistributedEnergySystemsin-China.pdf (Accessed 15 October 2018).

IRENA, 2018. Renewable energy and jobs. Annual review 2018. Report prepared by the International Renewable Energy Agency (IRENA). Available at: www.irena.org/publications/2018/May/Renewable-Energy-and-Jobs-Annual-Review-2018.

Kirkegaard, J.K., 2019. *Wind Power in China. Ambiguous Winds of Change in China's Energy Market*. Abingdon & New York: Routledge.

Koch, A., Girard, S. and McKoen, K., 2012. Towards a neighbourhood scale for low-or zero-carbon building projects. *Building Research & Information*, 40(4), pp. 527–537.

Korsnes, M., Berker, T., and Woods, R., 2018. Domestication, acceptance and zero emission ambitions: Insights from a mixed method, experimental research design in a Norwegian Living Lab. *Energy Research & Social Science*, *39*, 226–233.

Kubli, M., Loock, M. and Wüstenhagen, R., 2018. The flexible prosumer: Measuring the willingness to co-create distributed flexibility. *Energy Policy*, 114, pp. 540–548.

Luo, G.L., Long, C.F., Wei, X. and Tang, W.J., 2016. Financing risks involved in distributed PV power generation in China and analysis of countermeasures. *Renewable and Sustainable Energy Reviews*, 63, pp. 93–101.

Marigo, N., 2007. The Chinese silicon photovoltaic industry and market: a critical review of trends and outlook. *Progress in Photovoltaics: Research and Applications*, 15(2), pp. 143–162.

NEA, 2014. 国家能源局关于进一步落实分布式光伏发电. (Further implementation of existing distributed PV policies). National Energy Administration (NEA). Available at: http://zfxxgk.nea.gov.cn/auto87/201409/t20140904_1837.htm (Accessed 14 May 2019).

NEO, 2018. NEO network member webinar, May 2018. Presentation by Alex Shoer, Seeder Energy.

Qi, T., Zhang, X. and Karplus, V.J., 2014. The energy and CO2 emissions impact of renewable energy development in China. *Energy Policy*, 68, pp. 60–69.

Tan-Mullins, M., Cheshmehzangi, A., Chien, S. and Xie, L., 2017. *Smart-Eco Cities in China: Trends and City Profiles 2016*. Exeter: University of Exeter (SMART-ECO Project).

Wallace, W.L., Jingming, L. and Shangbin, G., 1998. The use of Photovoltaics for rural electrification in northwestern China. National Renewable Energy Laboratory, NREL/CP-520-23920, presented at the 2nd World Conference and Exhibition on Photovoltaic Solar Energy Conversion; 6–10 July 1998; Vienna, Austria.

Williams, A., 2017. *China's Urban Revolution: Understanding Chinese Eco-Cities*. Bloomsbury Publishing. Kindle Edition.

Wu, J., Zuidema, C. and Gugerell, K., 2018. Experimenting with decentralized energy governance in China: The case of New Energy Demonstration City program. *Journal of Cleaner Production*, 189, pp. 830–838.

Xue, J., 2017. Photovoltaic agriculture—New opportunity for photovoltaic applications in China. *Renewable and Sustainable Energy Reviews*, 73, pp. 1–9.

Yuan, J., Sun, S., Zhang, W. and Xiong, M., 2014. The economy of distributed PV in China. *Energy*, 78, pp. 939–949.

Zhang, F., Deng, H., Margolis, R. and Su, J., 2015. Analysis of distributed-generation photovoltaic deployment, installation time and cost, market barriers, and policies in China. *Energy Policy*, 81, pp. 43–55.

Zhang, S., 2016. Analysis of DSPV (distributed solar PV) power policy in China. *Energy*, 98, pp. 92–100.

Zhang, S., Andrews-Speed, P., and Li, S., 2018. To what extent will China's ongoing electricity market reforms assist the integration of renewable energy? *Energy Policy*, 114, pp. 165–172.

8 Creating a world after whose image?

> The bourgeoisie, by the rapid improvement of all instruments of production, by the immensely facilitated means of communication, draws all, even the most barbarian nations into civilisation. The cheap prices of commodities are the heavy artillery with which it batters down all Chinese walls, with which it forces the barbarians' intensely obstinate hatred of foreigners to capitulate. It compels all nations, on pain of extinction, to adopt the bourgeois mode of production; it compels them to introduce what it calls civilisation into their midst, i.e., to become bourgeois themselves. In one word, it creates a world after its own image.
>
> (Marx and Engels (1848), *The Communist Manifesto*)

Marx and Engels' manifesto has an air of urgency and drama to it. Although it has been more than 170 years since it was written, with certain amendments we might agree that the above quote has some truth to it; today, a capitalist (or bourgeoisie) world order prevails. In this world order, it appears, in spite of 70 years of Communist Party dominance, China is no exception. Book titles, such as *Playing Our Game: Why China's Rise Doesn't Threaten the West* (Steinfeld 2010) and *China Goes Global: The Partial Power* (Shambaugh 2013), and *Will China Save the Planet?* (Finamore 2018), indicate that China has entered the world stage and along with this entrance follow uncertainties, expectations and even hope. But were Marx and Engels correct in their predictions that capitalism would take over the world? Has capitalism created a China after its own image, and has China been a mere adopter? Has China been 'battered down' and forced to capitulate? If we look further back in history, we would perhaps agree that the four famous Chinese inventions—gunpowder, paper, the compass and the art of printing—were not forced upon other nations. In the same vein, this book suggest that something more has been going on than simply 'receiving' and 'being forced' technology, or even having technology 'transferred'. China appears to be paving its own road, attempting to set terms for the journey ahead. In this journey, behind some of the more visible 'toll stations' and 'border controls', a negotiation of and struggle for influence is taking place. These struggles within the wind and solar industries in China are what this book attempts to better understand.

In this final chapter, I will summarise and discuss some of the findings of this book related to the question: How has China's wind and solar energy transition unfolded, and how can this process be understood and conceptualised?

Chinese institutions of technological learning—'breakolage'?

Chapter 3 showed that various governance practices and institutions were important in developing China's wind and solar power industries, beyond formal laws and regulations. In relatively different ways in the wind and solar PV industry, the Chinese government engaged in a process of policy learning characterised by fluctuations between fragmentation and centralisation. At the beginning of the onshore wind industry development, local governments were, for instance, allowed to approve wind power projects, but as the development took off, approval was centralised. Nevertheless, the centralisation process was not simply changed from one day to the next. Rather, the Chinese Communist Party (CCP) orchestrated a 'campaign' that also involved the state-controlled media, the career prospects of firm managers, the lending allowances for banks and the leashing of large state-owned enterprises (SOEs). Policy learning was also characterised by experiments which were scaled up as wind power prices and internal policymaking processes were fine-tuned. This speaks to the Chinese government's ability to experiment and learn in order to get policies right and to constantly adapt to new situations.

A complex set of interactions between government and industry actors defined the development of China's wind and solar industries through channels as wide ranging as the media, the personal career of managers, local government interventions, new business models and innovation from private and public companies. This type of technology development is similar to the 'bricolage' story presented by Garud and Karnøe (2003), as introduced in Chapter 2, but has some differences. Rather than a modest and steady micro-learning process co-shaped by several involved actors, the Chinese wind industry example exhibits a massive centralised effort to set up experiments and testing sites which catered for steep learning curves and rapid development. It is tempting to call it a 'breakolage' approach that combined breakthrough with learning on several levels. Policy learning in this respect represented a type of reflexive governance where policymakers showed an effort and ability to adapt to and shape the rapid pace of changes within the wind industry. The developments of China's offshore wind industry and the solar PV industries also fit into this picture to some extent. For instance, it is possible to argue that the fragmented nature of China's governance-system is what allowed the solar PV industry to gain a foothold in the first place, as central government stakeholders were not necessarily interested in the industry in the early 2000s (Binz and Anadon 2018). As we saw in Chapter 4, local governments were convinced, and slowly but surely catered for solar PV manufacturing. Still, as soon as international markets started failing around 2009, the central government started working towards putting in place a domestic market (Andrews-Speed and Zhang 2019).

Chinese firms were also seeking to produce offshore wind turbines which corresponded to the quality of turbines from European firms. In this way, European heuristics, expectations and practices to some extent influenced Chinese firms. However, as presented in Chapter 6, certificates of quality conformity were unimportant in terms of actual procedures but made a difference in terms of 'face value' that gave companies with foreign certificates an edge. Since Chinese firms saw the European standard as a quality benchmark, turbine manufacturers were able to circumvent the time-consuming certification process by opting for 'quasi-certificates' that were satisfactory to the project developers. To Chinese firms in the offshore wind industry, certificates were therefore as much status symbols as they were important for improving product quality. As pointed out in Chapter 5, European certification and advisory companies made an effort to enrol Chinese companies into their criteria of product performance and certification. The established standard—and with it, the quality-agenda—was therefore not only 'transmitted' to Chinese firms, but certification and advisory agencies were also 'pushing' it on Chinese firms.

Chapter 5 argued that local employees of European companies are better at assessing the situation at hand than are European managers, as the local employees try to adapt the European services to local needs. However, the Chinese industry resisted the efforts of European companies to become 'obligatory passage points' in the process of developing offshore wind. Chinese companies apparently wanted to walk the path alone in order to ensure their own learning. The insights from Chapter 5 are telling in terms of how we should look at catch-up: instead of Chinese firms catching up with European firms and European firms 'providing' technology, we may rather observe an attempted 'interessement' and 'enrolment' of Chinese actors by European ones. In other words, standards, certification and advisory agencies assume an important role in innovation processes in Europe. However, in the context of China's offshore wind industry, standards and certificates are ambiguous and highly political instruments which are not always carried out in the 'great spirit' of innovation. Hence, instead of identifying the Chinese lack of interest in certificates as a 'failure' on the part of Chinese offshore wind companies, we may understand it as a struggle for independence. Altogether, this means that learning efforts of Chinese firms must not only involve the acquisition of 'tacit' and 'codified' types of knowledge but must also make assessments of what knowledge is necessary for their purpose and what knowledge is not. This may also be one of the large differences between the wind and the solar PV industries. In a recent comparison between the international knowledge flow of China's wind and solar PV industries, Zhou, Pan and Urban (2018) found that the wind industry has a stronger tacit (e.g., personal interaction between engineers and managers) knowledge flow, whilst the PV industry has more explicit (e.g., patent citations) knowledge flow internationally. The natures of the two technologies are different, which is also why different forms of learning mechanisms have been used (Zhou, Pan and Urban 2018).

There is also something to say for the various ways in which local Chinese companies have been preferred, but still within the regulations of the World Trade Organization. Tenders are released in Chinese, and documentation, permits, contracts and project design are negotiated locally (Beaumont 2019). Thus, in many cases there are local policy processes which can only by dealt with by Chinese firms. What is more, uncertainty about the future can best be dealt with by larger firms which have the financial means to allow capacities to be idle over time. In effect, this means that SOEs are particularly fit to survive in the Chinese environment. In addition, SOEs have ample access to resources, opportunities to diversify into new industries and have important government connections. They support local suppliers and are generally able to draw on competencies acquired in related industries in different branches of the company. With all their flaws, SOEs can be seen as important nodes for institutional change in China (Nolan and Wang 1999), and hence, important to support a renewable energy transition.

Overall, the above accounts demonstrate that, as we go into more detail of studying the development of the wind and solar industries and the strategies of the various actors involved, two points can be made. First, China is not copying an existing system, but attempting to create something independent, and second, innovation in a European context is not the same as innovation in a Chinese context. From a European point of view, these two points are related because as European companies understand successful practices through their previous experience and evaluate a potential Chinese success through the same criteria (i.e., they fear that Chinese companies will copy them), they also assume that innovation is the same in China as it is in a European context.

In order to survive in a global business environment, Chinese companies strive to create windows of opportunity which make them special. Nahm (2014, p. 186) argues that Chinese solar and wind firms have not

> emulated technological leaders but have specialized in innovative capabilities unique to Chinese firms. These contributions of Chinese wind and solar firms to the development and commercialization of high-technology products in global production networks break with patterns set by firms in other East Asian economies, which long produced cheap replicas of established products before catching up to the technological leaders in the West.

Nahm (2014) therefore proposes that Chinese companies have contributed unique opportunities for commercialisation that have benefitted the global wind and solar industries, and that the strategy of Chinese companies can be described as performing ‘innovative manufacturing’. Hence, we may look at

Chinese catching-up processes as contributions which add something new, and link existing knowledge and practices in novel ways. This has also been pointed out by several recent studies (Nahm and Steinfeld 2014; Rein 2014; Vinig and Bossink 2015). Based on my own analysis and these accounts, I therefore propose that we start to understand the catching-up process as a learning process which engenders unforeseen opportunities, wherein much negotiation takes place according to a multitude of interests.

To summarise, using European innovation 'boxes' for Chinese innovation phenomena is only to a certain extent suitable to understand and adequately conceptualise innovation in China's wind and solar industries. We find that Chinese companies are attempting to do things independently from European firms, and the Chinese government is not only attempting to accommodate for the learning processes of Chinese firms. Rather, the government is also adapting and learning, similar to how transition management theory describes 'reflexive governance', as introduced in Chapter 2. The Chinese government does not attempt to maximise control and attain a complete knowledge of the field, but is geared towards continued learning as events develop. This type of reflexive governance is what we may regard as policy learning, which distinguishes the Chinese approach from characterisations within innovation system perspectives. Does this mean that the Chinese government is catering for renewable energy industries according to the claims of the 'sustainability transitions' literature? In the next section, I will look more closely at the way in which some of the tenets of the sustainability transitions literature resonate with the Chinese experience.

Experiments, niches and learning

China's emerging wind and solar industries could be understood through the lens of 'strategic niche management', as introduced in Chapter 2, which would see them as niches which needed to be protected and nurtured in order to grow. This protection could, for instance, be in the form of a government subsidy—although for the solar PV industry this would actually include German feed-in tariffs nurturing the Chinese PV industry. History shows that state support is crucial in the nursing phase of a new technology's development (Kemp 1994; Jacobsson 2011). An illustrative example can be taken from the United States, where nuclear and fossil energy technologies historically received larger government subsidies than renewable energy technologies:

> the federal commitment to [oil and gas] was five times greater than the federal commitment to renewables during the first 15 years of each subsidies' life, and it was more than 10 times greater for nuclear.
>
> (Pfund and Healey 2011, p. 6)

More generally, Smith and Raven (2012) find that, in transition processes, protection of niches has three distinct (but simultaneous) properties: shielding, nurturing and empowerment. Shielding and nurturing are important in most accounts of strategic niche management, and Smith and Raven's (2012) contribution is on the aspect of empowerment: how actors create new associations and make a larger number of people and groups interested in the technology at hand—similar to actor-network theory's (ANT) description of translation and innovation.

The fourth, fifth, sixth and seventh chapters of this book have shown the various ways in which (formal and informal) institutions, actors and practices have shielded and nurtured onshore and offshore wind turbine development, as well as the solar PV industry. From this it was clear that Chinese companies wanted to take more action and be an active part in the innovation process, and not only rely on European procedures. This can be interpreted as another form of safeguarding domestic interests. In Chapter 6, we saw that a protective space was created in China's offshore wind industry, as exemplified by the ambitious government development targets which have been set. It appears, however, that the dynamics of niche management are different in this (catching-up) context. Rather than filling a nurturing and empowering role, state support in China's offshore wind industry was seemingly important, but companies were also expected to learn on their own. Thus, although a protective space for offshore wind technology was created and shielded, nurturing and empowering was instead left open and up for definition. This is the background for the seven-year delay which the industry experienced between the initiation in 2010, and take-off in 2017. This is what Chapter 6 called a 'strategic waiting game', which is characterised by ambitious government intentions but ambiguous signals of implementation. This waiting game appears to have a central function: reducing uncertainty about *direction* and *speed* of change. The central government surely does not have all the facts on the table and does not support a new industry unconditionally. Likewise, local industry and SOEs cannot cover all the costs themselves, and therefore stall.

Approaches within sustainability transition studies, such as the multi-level perspective (MLP), define actors which are operating within a sociotechnical regime. However, in some empirical studies, it has proven difficult to distinguish actors as being confined uniquely to the realm of the concepts 'niche', 'sociotechnical regime' and 'landscape' (Shove and Walker 2007; Jørgensen 2012; Åm 2015). This book has also highlighted that 'niche' actors in China are, to a great extent, established SOEs, often with experience in fossil fuel industries. In other words, they may be classified as actors from an established sociotechnical regime. As pointed out in Chapter 6, the Chinese government and industry functioned as much as collaborators in offshore wind technology empowerment as competitors. The government is interested in sustaining the growth of offshore wind technology and established state-owned industry actors will not actively counter this proposition. The industry is concerned about the details of how projects should be carried out and provides checks and balances on government visions and ambitions. In other words, it is not only that small offshore wind niches engage in active enrolment of the established actors, but that 'regime' actors are also actively involved in enrolling niche actors.

Rather than evaluating Chinese industry actors within the wind and solar PV segment on basis of their potential status as 'protected spaces' or 'niches', it might make more sense to look at the ways wind and solar industry actors approached the industry. In the interview data underlying this book, there was a general consensus amongst both Chinese and foreign actors that Chinese wind and solar PV firms 'take the bull by the horns' with an attitude of trying first and improving afterwards—they 'first do it, then solve it' (Kirkegaard 2015, p. 372). To use a Norwegian expression, they attempt to build a boat whilst at sea. The Chinese approach is related to their use of experimentation as a strategic procedure to learn and fine-tune policy, as has been pointed out by other scholars (e.g., Heilmann 2008a; 2008b; Kirkegaard 2019). This book has shown that experimentation was an important avenue for policy learning in China's wind and solar industries. Moreover, Chinese industry stakeholders—who often ignored foreign service providers with knowledge and experience in offshore wind technology—above all wanted to gain firsthand experience. Simply put, European service providers which wanted to manage risk did not succeed in China because Chinese firms were not interested in managing 'European' risks. Chinese wind and solar firms were interested in learning by conducting own experiments.

Indeed, it appears as though the Chinese government's management approach that allowed for experimentation created problems more quickly, and thereby induced learning to happen more quickly, as well. Kirkegaard (2015, p. 373) demonstrates this 'appropriate lack of planning' in China's wind industry, and argues that through this lack of planning, 'China has been forced to be innovative in solving them [problems which occur], and thus to learn faster'. The Chinese government, in other words, laid the foundation for these strategic emerging industries, but did not deal directly with the nitty-gritty of exactly how the road ahead should look. If we return to Latour's concepts 'matters of fact' and 'matters of concern', we may say that the CCP set a direction which appeared as a matter of concern. As this was done, the state allowed industries to move in a desirable direction, and expected matters of fact to appear later. In a European context, an 'appropriate lack of planning' would be deemed a highly risky approach, but, in China, the risk was considered part of a larger learning experience. This is not to say that Chinese firms blindly plunge into new projects without an idea of how it could work and be profitable. Rather, there is an interest in trying things out, and dealing with problems as they occur—leaving room for improvisation. Hence, a protected space where all perceived risks were identified and attempted catered for would take too long time to set up and perhaps also limit learning opportunities.

In general, we may conclude that the distinction made between levels within sustainability transition studies makes less sense in a Chinese context, since forces moving towards a 'new' regime originate as much from within the regime as from within 'niches'. The role of SOEs is especially important in this respect, given their size and amount of support and involvement with the Chinese wind and solar industries. Perhaps better names for the process within which the wind and solar industries have flourished are 'development arenas' (Jørgensen and Sørensen 1999; Jørgensen 2012), 'experimental points' (Heilmann 2008a; 2008b), 'sustainability experiments' (Berkhout *et al.* 2010; Berkhout, Wieczorek and Raven 2011) or simply 'strategic emerging industries', as the Chinese government itself emphasised in the twelfth Five-Year Plan (Bound *et al.* 2013). No matter what we call it, it can be described as a prioritised arena which allows for experimentation, but with efforts made to learn quickly and build new strategic industries to provide electricity and produce economic growth.

In the next section I take a closer look at the 'arts of the state' in China, and whether these can be described as state of the art with respect to accommodating new ventures and developing China's wind and solar PV industries.

The arts of the state

In examples of technological catching-up processes from Japan, Korea and Taiwan, the role of the state is evident to such an extent that the term 'developmental state' has been coined to explain their experiences (Amsden 1989; Stubbs 2009). More recently, the importance of the government in developed countries has surfaced. Mazzucato's (2013) book, *The Entrepreneurial State*, expresses the importance of the state to pave the way for private business in innovation processes. As she points out:

> when organized effectively, the State's hand is firm but not heavy, providing the vision and the dynamic push (as well as some 'nudges'—though nudges don't get you the IT revolution of the past, nor the green revolution today) to make things happen that otherwise would not have. Such actions are meant to increase the courage of private business.
>
> (Mazzucato 2013, Kindle loc. 495–496)

To Mazzucato, an 'entrepreneurial state' is seen as a steward that facilitates innovation. The concept of the 'entrepreneurial state' has recently been adapted to China, where it is affirmed that the government assumes an 'entrepreneurial' (i.e., catering for technology innovation) rather than 'developmental' (i.e., catering for manufacturing-led growth) role (Sun 2015). In the following, I examine the practices which constitute what I have termed the 'arts of the state', and then discuss how these compare with 'transitions management' and other concepts which address the role of government in industry development processes.

The Chinese government caters for new ventures on the one hand and, on the other hand, thinks about long-term plans. This not only fits well with Mazzucato's description of an 'entrepreneurial state', but also with the notion of 'transition management' that combines elements of long-term planning with incrementalism. As identified by Smith, Stirling and Berkhout (2005,

p. 1506), visions of the future play an important role in transition management literature, as visions fulfil at least five different functions: (1) they create a 'possibility space'; (2) they can act as 'heuristics'; (3) they are 'stable frame[s] for target-setting and monitoring'; (4) they can be metaphors for 'building actor-networks'; and (5) they can be 'narrative[s] for focusing capital and other resources'. Such visions are similar to what Chapter 6 defined as 'sociotechnical imaginaries'.

The development of China's onshore industry was facilitated by clever statesmanship by the CCP, entailing policy experimentation, implementation and learning. Some characteristics of China's energy governance cater for protracted decision-making, but fragmentation of authority may, in some cases, be beneficial in order to escape deadlock. However, policy learning alone does not sufficiently describe how movement into new initiatives is enabled; mobilisation of expectations and visions is also important for technologies to succeed. The Chinese government, by drawing on the future as a resource for change, was able to orchestrate the development of offshore wind technology, in spite of expectations of high costs and protracted development procedures. Chapter 6 addressed this topic and scrutinised the way in which the CCP, through negotiation with industry actors, was able to mobilise the future into present-day action. Indeed, looking into the future in order to assess its potential development trajectory is a craft the Communist Party has perfected since the first Five-Year Plan was launched in 1953 (Heilmann 2008a).

Chapter 6 found that the Chinese government had quite ambitious development targets for the future but remained rather ambiguous in relation to concrete support schemes for offshore wind. Ambiguity and uncertainty surely exist in other countries with ambitions to develop offshore wind technology—for instance, Norway (Heidenreich 2014). There are, however, several interesting differences between Norway and China in this respect. On the one hand, although uncertainty and ambiguity exist in China, industry actors are certain that things will happen 'soon'. Offshore wind has become a topic for the very top of the government; involved actors attempt to position themselves before 'take-off'. In Norway, on the other hand, although several firms within offshore oil and gas are interested in offshore wind, the commitment from the government is unstable and rather unsupportive (Normann 2015; Hansen and Steen 2015). In China, therefore, imaginaries set out by the government function as loose planning tools which are relatively concrete but still leave room for negotiation, whereas in Norway, plans are vague and change frequently—often fluctuating with a new minister or a new government. Chinese firms would be certain that something would happen soon and were not dwelling on questions of 'if' it would happen, but rather 'how soon' and 'how much'. Hence, offshore wind technology in China is not only a 'protected space', but also an 'expected space'.

Although it appears that the 'niche' concept does not fit with the Chinese situation, transition management theory and the reflexive governance concept resonate rather well with the way in which the Chinese Communist

Party was able to accommodate for wind turbine and solar PV technology. Transition management has developed its conceptualisation with the intention of advising governments on managing a transition away from a carbon-intense economy (Rotmans, Kemp and van Asselt 2001). But transition management portrays a rather idealised picture of how experimentation is driven by joint problem-solving, and it disregards how politics can be a nasty process dominated by powerful actors and their interests (Voß and Bornemann 2011). Transition managers are not neutral and do not necessarily represent sustainable projects (Shove and Walker 2007). Picking technologies and supporting certain actors will have political implications. In the case of China's offshore wind industry we might on the one hand say it is easier for latecomers to pick winners, as latecomers have more knowledge of how well an emerging technology may work, and thus opting for renewable energy technologies only after seeing that they can be profitable. On the other hand, there are several arguments which speak to the CCP's interests in keeping the status quo; the government has a strong interest in maintaining economic growth, providing enough electricity for its citizens and ensuring the continued legitimacy of the Party. In this consideration, it is also very likely that sustainability, in terms of reducing the pollution around major cities and reducing environmental impacts from coal power, plays an important role in the motivations behind the Chinese government's support for wind and solar technology (Ydersbond and Korsnes 2016). Moreover, it is certainly not a risk-free endeavour to develop these renewable energy technologies. Hence, there is a heterogeneous set of concerns and conflicts of interest that impact the decisions of policymakers and other industry actors.

Transition management and reflexive governance cover some important aspects which help us understand how policy learning may occur through emergent and experimental processes: if a project fails the first time, we may take that turn of events into account and learn from it in the next round. Transition management could therefore work as a practical and applicable guideline or set of advice for policy makers. However, it cannot be said to be a framework which helps us conceptualising innovation in China's wind and solar PV industries. This is particularly so, since the transitions management framework is founded a multilevel designation of niches, regimes and landscape (Rotmans, Kemp and van Asselt 2001). As already discussed, these levels do not resonate as well with the situation described in China. With actor-network theory (ANT) the focus would not be on preexisting levels, but we would rather have to ask: Who are the actors behind transition and what are their interests?

There is no shortage of scholars attempting to coin a term to explain China's transition over the past years; proposed terms include 'authoritarian upgrading' (Heilmann 2010), 'Janus-faced state-led growth' (Hsu 2011), 'crony communism' (Dickson 2011) and 'structured improvisation' (Ang 2016). Most concepts hint at the important role of the state as well as the ambiguous but adaptive nature of the Chinese governance apparatus. Chinese actors are attempting to create an arena wherein wind and solar PV technology is a favourable option alongside other cheap electricity providing technologies. To create this arena, a massive work of translation of a large number of actors, such as private and state-owned companies, local, provincial and central governments and foreign actors, must be aligned with their expectations of a future renewable energy industry in China. This work is what Chinese actors have started by creating space for a new influx of ideas, actors and competencies within wind and solar PV technology. Instead of coining new concepts, perhaps the best we can do is to describe the 'arts of the state' and the ways in which these strategies pave the way for new initiatives in China.

The 'arts of the state' in China are unique and Chinese institutions do not hinder innovation. To describe the development in the words of ANT, the state acts as a strong agent of translation in favour of new renewable energies. In effect, several actors try to translate their interests and goals in order to align with others: the state, with vision and long-term ambition, the industry, with a more immediate and localised concern, and European advisory and certification agencies entering China, with their own ideas of how wind and solar PV technology should be practiced. Andrews-Speed and Zhang (2019, p. 271) recently concluded the following about China's renewable energy success:

> In short, our analysis shows that the sources of success lie, to a great degree, in the deployment of capital in three forms: political capital deployed by the top leadership to develop a strategic vision and to follow through robustly by means of its substantial state capacity; financial capital in many forms provided by the central and local governments and the state-owned banks; and human capital represented by large numbers of dynamic, entrepreneurial and technically trained individuals in state-owned and private enterprises across the energy sector—all embedded in a huge domestic market.

This sounds like a very robust conclusion, but it also smoothes over some of the more difficult negotiation processes between the political, the financial and the human capital. These negotiation processes, I argue, are not negative externalities to the Chinese energy governance system, but embedded and central to achieving the results which have been achieved. As Chinese actors attempt to cater for innovation, they also shape the environment that constitutes innovation. This brings us to the role of controversy in innovation processes within China's wind and solar PV industries.

This book has illustrated how different actors attempt to bring their own perspectives, needs and expectations into play in China's wind and solar industry development. The Chinese government, the private enterprises and SOEs from various industry segments, European certification and advisory agencies, and international actors all have their idea of what they want from the industries and how the industries should develop. All these actors work to achieve some form of stability or settlement where their version of the world is represented. In other words, all these actors are part of controversies around China's wind and solar industry development. They all impact the way in which the industries develop by engaging in this controversy, although they might not in the end succeed with their own objective.

One way of understanding this is through the concept of assemblages, as introduced in Chapter 2. The efforts of the various actors may be attempts at associating and assembling people, things and concepts which align with their own interests and objectives. As we have seen, the Chinese government actively encouraged domestic wind companies to engage with the industry but was also reluctant to enrol European companies to a great extent. Competencies which were mobilised were either distinctly Chinese, or were already 'Chinesified', so that full ownership and credit could be given to Chinese actors. In this sense, we may talk about a form of 'Chinesification' assemblage where competencies, national values, standards, experiments and risks were translated into wind and solar PV technology development projects. A similar strategy could, for instance, be found in the Norwegian oil and gas sector in the 1960 and '70s, when the government played an important role not only in establishing the Norwegian company Statoil (now 'Equinor'), but also in catering for research, education and demonstration, and adapting technology to Norwegian culture and environment (Sørensen 2005). Briefly put, Norwegian actors took oil and gas technology and experiences from the United States and the United Kingdom and adapted them to Norwegian conditions, leading to several innovations, such as the Condeep structure, sub-sea installations and horizontal drilling, which made the Norwegian experience special (Sørensen 2005). What is more, we cannot say that US and UK companies necessarily were negatively impacted by the Norwegian experience on a longer term. In China, we may well observe that the act of rendering technology Chinese, or 'Chinesification', can be part of a larger innovation strategy.

Let's picture a 'competing' European wind and solar PV assemblage which is enacted in Europe and attempted 're-enacted' in China with actors such as the European certification and advisory agencies. Such an exercise takes us to the question of what type of knowledge and experience are really *enacted*; what are the important types of knowledge which are displaced and enacted in China's wind and solar PV industries? We have seen that, what in a European context was termed risky and reckless (i.e., 'they just get on and do it'), low-quality, too ambitious or arrogant could in the Chinese case often be characterised as a difference in priority; a different appraisal of what knowledge to be enacted when. Chinese actors tended to see the European approach as overengineered, overprotected and perhaps even somewhat arrogant. Simply put, if the whole European package would be accepted, several European actors would be attached to it, and that would be unwanted if independence is to be achieved for Chinese actors. For this reason, 'rendering Chinese' is a much more long-term strategy than a simpler 'copy and paste' strategy. The same way the international oil-and-gas giant Equinor would not exists had Norway simply let US and UK actors do as they wanted, China aims at establishing their own domestic and independent industry. To be sure, it is not a given that such a strategy would succeed, nor that it is a great idea. We should also not say that Chinese actors deliberately produce low-quality products or want to be reckless. Rather, the strategies chosen reflect a different point of view and a different goal, which to European eyes appear risky, but to Chinese actors appear necessary.

Since ANT focuses on controversy and does not take an assumed view of the world as a point of departure, it allows for categories to be redefined in the eyes of the beholder. ANT points out that there are certain downsides in simplifying how we view innovation, and this may be especially true in a new context, such as in China. Thus, ANT enables a better understanding of the Chinese side of the story, and why Chinese actors choose the strategies they do. Innovation in China's wind and solar PV industries may in this light be described as a practice which draws from and builds on assemblages—a process of intentional development through iterative and pragmatic trial and error. This is illustrated by the specific way the Chinese central, provincial and local government interact with the wind and solar PV industries, the way which experiments and risk-taking is catered for in ways which go beyond what European companies would consider necessary, the way expectations and visions are managed and aligned, the special role of SOEs within the wind industry, and the impeccable timing of assisting the solar PV industry.

Intermittency, flexibility and deregulation

Electricity grid reform is no simple task either in China or other places in the world (Weston and Dupuy 2019). China has already gone through an impressive electric system reform since the 1980s, and a system which used to be plagued by blackouts has now become the largest and most technologically sophisticated in the world, providing stable electricity to almost one fifth of the world population (Weston and Dupuy 2019). Deregulation and introducing market-based principles of electricity is not straightforward. Electricity markets do not just appear out of nowhere. Rather, they are carefully crafted, and 'market agents perpetually bargain the terms of exchange, negotiating the price, quantity, and quality of goods, but also work to shape the very organization of markets' (Blanchet and Depeyre 2016, cited in Kirkegaard and Caliskan 2019, p. 4). Questions of market reform are once again on the table, also in Europe, after the introduction of more intermittent renewable energy technologies which challenge the way in which electricity is managed (Pollitt and Haney 2013). In many ways, the introduction of wind and solar power to an electricity grid forces a change from a more static load management to a more active load management. It is relatively easy to predict and control the electricity production of a coal power plant, but this becomes trickier with wind and solar energy.

For China, it appears that finding the right balance between giving up and keeping control is the main difficulty that is faced in the coming reform process. Providing cheap and stable electricity to the population is a central concern to the CCP. The Chinese government has recognised the highly inefficient way the current system is working and realises that reform might be a good way to reduce wasted electricity. Indeed, the priority which has been given to coal-fired power station in electricity dispatch is one of the primary reasons behind the large curtailment of solar and wind power (Kahrl, Dupuy and Xuan 2016; Pollitt, Yang and Chen 2017). And local governments and coal-fired power station do not want to give up their position easily. A recent report by Agora Energiewende (2018, p. 23/24) explains the difficulty:

> Because demand is not growing at the same pace as capacity, the projected share of coal power in 2020 significantly overshoots the economical optimum. In a competitive market, operational hours would decrease significantly, shrinking revenues accordingly. It is easy to imagine that the owners of coal plants are not the most enthusiastic proponents of the economic dispatch model. Local governments, whose coal assets tend to be smaller and less efficient, are particularly afraid of such a change.

As elaborated on in Chapter 4, the large mismatch between generation and distribution capacity is related to the fragmented nature of China's energy governance. Local and provincial governments were eager to generate local economic growth, large SOEs needed to fulfil certain criteria related to renewable energy targets, and these then push forwards with project construction although grid infrastructure is not yet in place (Luo *et al.* 2018). What is more, although SOEs might end up losing money in the short run on a wind farm due to curtailment, in the long run developing a project would still make sense because they have gained (temporary) ownership of a valuable wind resource site. As pointed out by Zhu *et al.* (2019, p. 200), developers 'competed fiercely over wind power sites, increasing investment scale as a rational long-term strategy of profitability'.

Chapter 7 showed that there has been an explosion in the number of distributed solar PV installations throughout China, with an especially rapid growth since 2015. This development implies something more than just an impressive addition of solar PV capacity. It implies that there are a number of new types of local business models for financing, projecting, installing and connecting local PV panels on rooftops and grounds in both urban and rural areas where people consume the electricity. This change has carried with it a new challenge to the local grid operators and electricity dispatch centres, which China is now in the middle of a process of solving. Given the intermittency of wind and solar power, we could ask if a country with an authoritarian governance system will be able to flexibly manage the grid system which is required? By 2020, the plan is to introduce more market-mechanisms to decide electricity prices, which may make it easier to sell and buy electricity generated from renewable energy sources (Zhang, Andrews-Speed and Li 2018).

As Chapter 7 illustrates, the development towards more decentralised electricity generation is not without its issues. State Grid emerged as an obstacle to some of the issues related to restructuring, and this will likely take time to change. This was for instance related to connecting PV panels to the grid, and to reveal the real costs of electricity transmission, something which only the grid company really knows. This means that the 'smart' or 'prosumer' agenda which is seen discussed in European countries is still several years into the future in China. It is also not a given that the Chinese 'smart' or 'prosumer' society will look anything like the one imagined in Europe. New types of business models are emerging, which cater for increased use of distributed renewable energy. In Europe, the emphasis is on increased energy efficiency through empowering the consumer. In Chapter 7, we saw in one of the cases which people did not really care much about where their electricity came from, as long as it worked—and they earned some extra income on having PV panels on their roofs. Perhaps China will develop a more decentralised system, which achieves a green electricity supply, but where supply and demand of electricity is solved through different means. For instance, there might be a room for 'aggregators' that take care of electricity balance through smart technologies and include batteries from China's massive electric vehicle fleet in a comprehensive smart system. It would not be surprising to see that China's electricity deregulation process starting from the year 2020 will include a wide variety of new business models, some of which will be entirely new, leapfrogging solutions used in Europe.

China's wind and solar energy transition: innovation and sustainability

> The future belongs to hybridisation rather than to clearly defined lines of development.
>
> (Akrich, Callon and Latour 2002, p. 212)

This last section summarises what has been achieved in this book and provides some reflections around wider implications of this analysis. Several recent contributions have described China as innovating and developing new technologies (e.g., Nee and Opper 2012; Nahm 2014; Wan *et al.* 2015; Tyfield 2017). However, with regard to China's ability to innovate and create new technology, there are 'believers' and 'doubters' (Steinfeld 2010), or 'optimists' and 'pessimists' (Tyfield 2017). Depending on the definition of innovation at hand and the sector chosen for analysis, China can be described as either lagging or leading. With the broadest definition of innovation, China is found to innovate simply by interacting and engaging with new technologies in a global context; with the stricter definition, China does not and cannot innovate because nothing genuinely new is created. In this book I have argued that innovation in this Chinese context is a process characterised by learning, negotiation and a fine balance between chaos and control. In a globalised world, it is naïve to believe that China will become fully 'independent' or able to conduct purely 'indigenous' innovation. However, it is equally misleading to think that China simply copies technology from foreign countries and that the learning process ends with that.

What we observe in China is a process of development that impacts the rest of the world. Thus, rather than talking of independent innovation, we could talk about interdependent innovation when describing China's development strategy. As convincingly pointed out by Ang (2016, Kindle loc. 4485): 'development is a coevolutionary process, not only in China but also in other parts of the world and in recent and historical periods'. The nuances of this interdependent innovation process need to be elaborated on and described more accurately, and without a preconfigured understanding of what innovation is. This book set out to present this kind of detailed analysis of the innovation processes, the trails and nitty-gritty aspects of the wind and solar energy transition in China. Thus, the main contribution of this book has been to provide a more detailed picture of where the locus of change can be said to be, what actors are important, *how* these actors have manoeuvred to become important and the ways in which learning takes place at multiple places simultaneously. China appears to be going in a direction that is deliberately chosen, and one that is planned, but still with an 'appropriate lack of planning' (Kirkegaard 2015, p. 373). The contribution of this book is a richer understanding of the innovation process of two emerging technologies—wind turbines and solar PV—in China.

By describing and drawing from approaches such as innovation systems, sustainability transitions and actor-network theory (ANT), this book contributes to understanding better the way in which change towards more sustainable technologies can be understood and conceptualised in China. Chapter 2 started out by critiquing the way in which technological learning has been conceptualised through 'technology transfer' and 'catch-up'. The 'catch-up' term implicitly contains several assumptions related to the experiences of northern Europe and the United States since the Industrial Revolution. If we focus on what has consistently been found amongst catching-up countries—namely that they *co-produce* new opportunities through a process of constant adaptation to changing conditions—we arrive at a relatively simple definition of catch-up: the knowledge that a potentially beneficial industry exists, and the learning process accompanied with this knowledge. This definition is not far from Schumpeter's definition of innovation as new combinations of existing knowledge.

Chinese wind and solar PV actors could choose to simply accept things as they have been defined by 'frontier' actors, but instead they choose to challenge them and take risks. This is perhaps most clearly the case with European certification and advisory agencies, but it is also a general finding throughout the book: Chinese actors take risks to learn. Naturally, many actors choose safe strategies aimed at rapid economic growth and shiny new factories through foreign direct investments (Xiao, Tylecote and Liu 2013), which can be termed 'copy and paste' strategies. Nevertheless, a general eagerness to engage in trial and error prevailed in many of the interviewee accounts in this book. Thus, I did not observe 'acidic attitudes towards the new and untried' as argued, for instance, by Baark (2007, p. 354), and which was elaborated in Chapter 3 by the 'pessimist' side to whether or not innovation occurs in China. Instead, Chinese actors appear to be avid learners, willing to take chances and quick to adapt and create new opportunities. This is defining of innovation in the Chinese wind and solar PV industries.

Instead of 'indigenous' innovation, we may conclude that certain elements of wind and solar PV technology development in China hint at a form of 'Chinesification' of technology which renders competencies and processes Chinese. Mazzucato (2013) defines entrepreneurs as actors who are willing and able to take on risk and genuine uncertainty. She describes the United States as a case wherein the state supported research, development and commercialisation of new technologies in very early phases (such as technologies which were used in the iPhone). The foresightedness, courage and availability of resources required to take on risks are central ingredients to state entrepreneurs. In the Chinese case, we observe strong state support of industrial innovation and manufacturing of technologies which already exist elsewhere. As we have seen throughout this book, the Chinese government and industry actors are also willing and able to take on risk and uncertainty. This risk is related to the uncertainty of rendering technologies and practices Chinese, and whether or not something new will come out of it in the end. In effect institutions and governance practices emerged concomitantly with the growth of China's wind and solar PV industries. In this way, innovation in China's wind and solar PV industries may be conceptualised and understood as a practice which draws from and builds on assemblages. Such assemblages are held together and shaped by the efforts of a heterogeneous set of actors with diverging aims, ambitions and interests. Experiments and risk-taking are catered for, expectations and visions are managed, negotiated and attempted aligned, and private and state-owned enterprises are productively engaged in ways which avoid micro-management and government overspending. Flexibility and adaptability allowed for iterative and continual learning. Hence, as described through comparing and contrasting the wind and solar PV industries, innovation in China can be considered a space where experimentation, learning and exchange of knowledge and experiences lead to a co-production of new, global opportunities.

Towards sustainability?

Having outlined the findings of this book, let's now reflect on the extent to which China's wind and solar industry developments represent a transition towards something which is more environmentally sustainable. One might, for instance, expect China's energy industry to be characterised by path dependency, as caused by a technological paradigm which influences learning processes. For instance, Unruh and Carrillo-Hermosilla (2006, p. 1189) make the case that 'carbon lock-in may become globalized and [...] large developing countries, if they are successful at rapid industrialization, will become carbon copies of their industrialized neighbors'. Although China, on the face of it, looks like a 'carbon copy' with large investments in coal, gas and oil, the nation is investing more in renewables than is any other country in the world: at the end of 2017 China invested more in renewable energy than the United States and Europe combined (REN21 2018). For this and more reasons, several authors have recently agreed that the world is moving towards a 'green revolution', to which China will contribute—and is already contributing—at the forefront (e.g., Mazzucato 2013; Hu 2014; Mathews and Tan 2015).

Depending on our definition of sustainability, we can discuss whether the wind and solar PV industry development is part of a larger transition to a more sustainable China. Focussing uniquely on energy supply, the scale of investments in nonfossil energy sources (including nuclear power) are now larger than those in fossil energy globally; more nonfossil electricity generation capacity is added each year, and nonfossil electricity power generation increases quickly, whilst fossil power generation decreased in 2014 (Mathews and Tan 2015). China is a central contributor to this, not only in terms of domestic installations, but also in contributing to reduce renewable energy prices globally (Green 2019). In the ten years between 2007 and 2017, despite increasing electricity consumption, coal's share in China's power sector has been reduced from 81 percent to 65 percent (Alva and Li 2018). These are indications that China is moving rapidly towards more nonfossil electricity sources. Moreover, the institutions, organisations and practices related to solar PV, onshore and offshore wind identified in this book are similar and relevant to other renewable energy technologies in China, such as biomass (Binz *et al.* 2017). Wind and solar energy, seen in this context, is part and parcel of a larger transition, and the success of these technologies in China may affect the success of other low-carbon energy sources, as well.

Nevertheless, since rapid growth comes with a price, caution should be exercised. China's economic turnaround, which increased incomes and the national gross domestic product, has involved decades of environmental abuse, destruction of limited natural resources and widespread pollution of air, soil and water (Shapiro 2012). In Chapter 4, it was indicated that the both the wind and solar PV industries once faced severe overcapacity issues, and the government had to tighten its leash on the industries. Such problems of excess capacity have not only been found in these two industries but appear to be a more general element in China's technological catching-up strategy (Rock and Toman 2015). For instance, the overcapacity in iron and steel industries in China in 2014 represented the total iron and steel capacity in Europe (Rock and Toman 2015, p. 250). This can also be seen in entirely different areas. As an example, in 2018 one fifth of all homes in China were empty—that equals to 50 million apartments (bloomberg.com 2018). What is more, there are several so-called ghost cities around China, basically empty apartment buildings waiting to be filled due to expected urbanisation. To build all these apartments, China has used a lot of concrete: in the three years between 2011 and 2013 China used more cement than the United States used during all of the twentieth century (bbc.com 2015). This strategy might lead to quick economic growth and substantial technological learning, but it also has several downsides, including a decreased focus on product quality, long-term performance and health, safety and environment (HSE), as addressed in the fourth, fifth and sixth chapters of this book.

Recently, the Intergovernmental Science-Policy Platform on Biodiversity and Ecosystem Services (IPBES) released a report summarising the detrimental effect human activities have on nature and Earth's ecosystems. Biodiversity and services which we derive from nature are in rapid decline, and these changes were found to be driven amongst other things by changes in land and sea use, climate change and pollution (IPBES 2019). If Chinese companies generate overcapacity in order to learn, and in these learning process waste large amounts of resources, the sustainability of China's renewable energy transition can be jeopardised. Some recent calculations made, show that the solar PV industry only recently has started contributing positively to the climate: 'despite the negative environmental effect of the PV industry in the initial stages (due to the heavily export-oriented strategy), stimulation of the domestic market resulted in effective CO2 reduction in the most recent stage' (Shubbak 2019, p. 1010).

In my view, there are still reasons to be optimistic about what is going on in China. Steinfeld (2010, p. 174) has pointed out that it is very likely that 'in an area like climate change remediation, technologies which prove indispensable globally will have had their origins in the Chinese market'. The velocity with which change is happening in China is transforming everyday life, policy practices and energy business models with consequences which go far beyond China. With the Belt and Road Initiative, China also has shown international ambitions for which solar PV and wind energy technologies will become important export-technologies. A static model or a preconfigured understanding of how the world looks cannot describe the multifacteted ways in which change takes place in China. China's trial-and-error approach is very concrete, gives quick feedback as to what works and what does not work, and could potentially represent a more progressive approach to innovation and technology development than the approaches employed elsewhere. Although such an approach can be clunky and inelegant, it may be able to respond more quickly to the increasingly negative impact we humans have on Earth's climate and environment. This book has attempted to go more into detail to unravel some of the complexities and intricacies which have underpinned change in China's wind and solar PV industries. This should be seen as a starting effort to open up discussions of change in China, rather than narrowing them down and fitting them into preconfigured boxes. The hope is that in this way can we be better prepared to be taken by surprise, and to some extent anticipate the ways in which things can go both terribly wrong and terrifically right.

References

Akrich, M., Callon, M. and Latour, B., 2002. The key to success in innovation part ii: the art of choosing good spokespersons. *International Journal of Innovation Management*, 6(2), pp. 207–225.

Alva, C.A.H. and Li, X, 2018. Power sector reform in China. An international perspective. Report published by the OECD/IEA, International Energy Agency.

Åm, H., 2015. The sun also rises in Norway: Solar scientists as transition actors. *Environmental Innovation and Societal Transitions*, 16, pp. 142–153.

Andrews-Speed, P. and Zhang, S., 2019. *China as a Global Clean Energy Champion. Lifting the Veil.* Singapore: Palgrave Macmillan.

Ang, Y.Y., 2016. *How China Escaped the Poverty Trap.* Ithaca, NY: Cornell University Press.

Agora Energiewende, 2018. A star for China's energy transition: five golden rules for an efficient transformation of China's energy system. A report by Agora Energiewende in collaboration with GIZ (Deutsche Gesellschaft für Internationale Zusammenarbeit).

Amsden, A.H., 1989. *Asias's Next Giant. South Korea and Late Industrialization.* New York, Oxford: Oxford University Press.

Baark, E., 2007. Knowledge and innovation in China: historical legacies and emerging institutions. *Asia Pacific Business Review*, 13(3), pp. 337–356.

bbc.com, 2015. Cement and pig consumption reveal China's huge changes. Available at: www.bbc.com/news/world-asia-china-33802777 (Accessed 29 May 2019).

Beaumont, H., 2019. Doing business in China's offshore wind market. Azure International. Presented at GWEC Webinar, 20 May 2019.

Berkhout, F., Verbong, G., Wieczorek, A.J., Raven, R., Lebel, L. and Bai, X., 2010. Sustainability experiments in Asia: innovations shaping alternative development pathways? *Environmental Science & Policy*, 13(4), pp. 261–271.

Berkhout, F., Wieczorek, A.J. and Raven, R., 2011. Avoiding environmental convergence: a possible role for sustainability experiments in latecomer countries? *International Journal of Institutions and Economies*, 3(2), pp. 367–385.

Binz, C. and Anadon, L.D., 2018. Unrelated diversification in latecomer contexts: Emergence of the Chinese solar photovoltaics industry. *Environmental Innovation and Societal Transitions*, 28, pp. 14–34.

Binz, C., Gosens, J., Hansen, T. and Hansen, U.E., 2017. Toward technology-sensitive catching-up policies: Insights from renewable energy in China. *World Development*, 96, pp. 418–437.

bloomberg.com, 2018. A fifth of China's homes are empty. That's 50 million apartments. 8 Nov. Available at: www.bloomberg.com/news/articles/2018-11-08/a-fifth-of-china-s-homes-are-empty-that-s-50-million-apartments (Accessed 22 February 2019).

Bound, K., Saunders, T., Wilsdon, J. and Adam, J., 2013. *China's Absorptive State. Research, Innovation and the Prospects for China-UK Collaboration.* Available at: www.nesta.org.uk/sites/default/files/chinas_absorptive_state_0.pdf.

Dickson, B.J., 2011. Who consents to the "Beijing consensus"?: Crony communism in China. In S.P. Hsu, S. Zhao and Y. Wu (eds) *In Search of China's Development Model: Beyond the Beijing Consensus.* New York: Routledge, pp. 189–204.

Finamore, B., 2018. *Will China Save the Planet?* Wiley. Kindle Edition.

Garud, R. and Karnøe, P., 2003. Bricolage versus breakthrough: distributed and embedded agency in technology entrepreneurship. *Research Policy*, 32(2), pp. 277–300.

Green, M.A. 2019. How did solar cells get so cheap?. *Joule*, 3(3), pp. 631–640.

Hansen, G.H. and Steen, M., 2015. Offshore oil and gas firms' involvement in offshore wind: Technological frames and undercurrents. *Environmental Innovation and Societal Transitions*, 17, pp. 1–14.

Heidenreich, S., 2014. *Blowing in the Wind: The Socialization of Offshore Wind Technology*. PhD Dissertation, Norwegian University of Science and Technology.

Heilmann, S., 2008a. From local experiments to national policy: the origins of China's distinctive policy process. *The China Journal*, 59, pp. 1–30.

Heilmann, S., 2008b. Policy experimentation in China's economic rise. *Studies in Comparative International Development (SCID)*, 43(1), pp. 1–26.

Heilmann, S., 2010. Chapter 6. Economic governance: authoritarian upgrading and innovative potential. In J. Fewsmith (ed.) *China Today, China Tomorrow: Domestic Politics, Economy, and Society*, pp. 109–129. Lanham, Md.: Rowman & Littlefield.

Hsu, S.P., 2011. In search of China's development model: beyond the Beijing Consensus. In S.P. Hsu, S. Zhao and Y. Wu (eds) *In Search of China's Development Model: Beyond the Beijing Consensus*. New York: Routledge, pp. 1–25.

Hu, A., 2014. *China: Innovative Green Development*. Heidelberg, New York, Dordrecht, London: Springer Berlin/Heidelberg.

IPBES, 2019. Media release: nature's dangerous decline 'unprecedented'; species extinction rates 'accelerating'. Available at: www.ipbes.net/news/Media-Release-Global-Assessment (Accessed 28 May 2019).

Jacobsson, S., 2011. Kapittel 2: Systembygging for ny energi. In J. Hanson, S. Kasa and O. Wicken (eds) *Energirikdommens paradokser: innovasjon som klimapolitikk og næringsutvikling*. Oslo: Universitetsforlaget.

Jørgensen, U., 2012. Mapping and navigating transitions—The multi-level perspective compared with arenas of development. *Research Policy*, 41(6), pp. 996–1010.

Jørgensen, U. and Sørensen, O.H., 1999. Arenas of development—a space populated by actor-worlds, artefacts, and surprises. *Technology Analysis & Strategic Management*, 11(3), pp. 409–429.

Kahrl, F., Dupuy, M. and Xuan, W., 2016. Issues in China power sector reform: generator dispatch. discussion paper. Beijing: The Regulatory Assistance Project. Available at: www.raponline.org/wp-content/uploads/2016/07/rap-kahrl-dupuy-wang-china-generator-dispatch-reform-july-2016.pdf (Accessed 26 May 2019).

Kemp, R., 1994. Technology and the transition to environmental sustainability: The problem of technological regime shifts. *Futures*, 26(10), pp. 1023–1046.

Kirkegaard, J.K., 2015. *Ambiguous Winds of Change—or Fighting against Windmills in Chinese Wind Power. A Constructivist Inquiry into China's Pragmatics of Green Marketisation. Mapping Controversies Over a Potential Turn to Quality in Chinese Wind Power.* PhD Monography, Department of Business and Politics, Copenhagen Business School. Available at: http://openarchive.cbs.dk/handle/10398/9116.

Kirkegaard, J.K., 2019. *Wind Power in China. Ambiguous Winds of Change in China's Energy Market*, Abingdon & New York: Routledge.

Kirkegaard, J.K. and Caliskan, K. 2019. When socialists marketize: the case of China's wind power market sector. *Journal of Cultural Economy*, 12(2), pp. 154–168.

Luo, G., Dan, E., Zhang, X. and Guo, Y., 2018. Why the wind curtailment of northwest China remains high. *Sustainability*, *10*(2), p. 570.

Marx, K. and Engels, F., 1848. *The Communist Manifesto*. Transcribed by Allen Lutins with assistance from Jim Tarzia. From the English edition of 1888, edited by Friedrich Engels.

Mathews, J.A. and Tan, H., 2015. China's green-energy revolution. *project syndicate*. Available at: www.project-syndicate.org/commentary/china-green-energy-revolution-by-john-a-mathews-and-hao-tan-2015-05 (Accessed 16 May 2015).

Mazzucato, M., 2013. *The Entrepreneurial State: Debunking Public vs. Private Myths in Risk and Innovation*, London, New York, Delhi: Kindle Edition.

Nahm, J. and Steinfeld, E.S., 2014. Scale-up nation: China's specialization in innovative manufacturing. *World Development*, 54, pp. 288–300.

Nahm, J.M., 2014. *Varieties of Innovation: The Creation of Wind and Solar Industries in China, Germany, and the United States*. Doctoral dissertation, Department of Political Science, Massachusetts Institute of Technology.

Nee, V. and Opper, S., 2012. *Capitalism from Below: Markets and Institutional Change in China*. Cambridge, MA: Harvard University Press.

Nolan, P., and Wang, X, 1999. Beyond privatization: Institutional innovation and growth in China's large state-owned enterprises. *World Development*, 27(1), pp. 169–200.

Normann, H.E., 2015. The role of politics in sustainable transitions: The rise and decline of offshore wind in Norway. *Environmental Innovation and Societal Transitions*, 15, pp. 180–193.

Pfund, N. and Healey, B., 2011. *What Would Jefferson Do? The Historical Role of Federal Subsidies in Shaping America's Energy Future*. Online: DBL Investors. Available at: www.dblinvestors.com/documents/What-Would-Jefferson-Do-Final-Version.pdf.

Pollitt, M., Yang, C. and Chen, H., 2017. Reforming the Chinese electricity supply sector: lessons from international experience. Cambridge Working Paper Economics: 1713.

Pollitt, M.G. and Haney, A.B., 2013. Dismantling a competitive electricity sector: The UK's electricity market reform. *The Electricity Journal*, 26(10), pp. 8–15.

Rein, S., 2014. *The End of Copycat China: The Rise of Creativity, Innovation, and Individualism in Asia*. Hoboken, New Jersey: Kindle Edition.

REN21, 2018. Renewables 2018 Global status report. Renewable Energy Policy Network for the 21st Century (REN21). Available at: www.ren21.net/status-of-renewables/global-status-report/ (Accessed 21 April 2019).

Rock, M.T. and Toman, M.A., 2015. *China's Technological Catch-Up Strategy: Industrial Development, Energy Efficiency, and CO2 Emissions*. Oxford, New York: Oxford University Press.

Rotmans, J., Kemp, R. and van Asselt, M., 2001. More evolution than revolution: transition management in public policy. *foresight*, 3(1), pp. 15–31.

Shambaugh, D., 2013. *China Goes Global: The Partial Power*. Oxford, New York: Oxford University Press.

Shapiro, J., 2012. *China's Environmental Challenges*. Cambridge, UK, and Malden, mA: Polity Press.

Shove, E. and Walker, G., 2007. CAUTION! Transitions ahead: politics, practice, and sustainable transition management. *Environment and Planning A*, 39(4), pp. 763–770.

Shubbak, M.H., 2019. The technological system of production and innovation: The case of photovoltaic technology in China. *Research Policy*, 48(4), pp. 993–1015.

Smith, A., Stirling, A. and Berkhout, F., 2005. The governance of sustainable socio-technical transitions. *Research Policy*, 34(10), pp. 1491–1510.

Smith, A. and Raven, R., 2012. What is protective space? Reconsidering niches in transitions to sustainability. *Research Policy*, 41(6), pp. 1025–1036.

Steinfeld, E.S., 2010. *Playing Our Game: Why China's Rise Doesn't Threaten the West.* Oxford: Oxford University Press.

Stubbs, R., 2009. What ever happened to the East Asian Developmental State? The unfolding debate. *The Pacific Review*, 22(1), pp. 1–22.

Sun, Z., 2015. Technology innovation and entrepreneurial state: the development of China's high-speed rail industry. *Technology Analysis & Strategic Management*, 27(6), pp. 646–659.

Sørensen, K.H., 2005. Jakten på innovasjon: Fra nasjonsbygging til dot.com. In Frønes and L. Kjølsrød (eds.). *Det norske samfunn*, pp. 511–535.

Tyfield, D., 2017. *Liberalism 2.0 and the Rise of China: Global Crisis, Innovation and Urban Mobility*. London and New York: Routledge.

Unruh, G.C. and Carrillo-Hermosilla, J., 2006. Globalizing carbon lock-in. *Energy Policy*, 34(10), pp. 1185–1197.

Vinig, T. and Bossink, B., 2015. China's indigenous innovation approach: the emergence of Chinese innovation theory? *Technology Analysis & Strategic Management*, 27(6), pp. 621–627.

Voß, J.P. and Bornemann, B., 2011. The politics of reflexive governance: Challenges for designing adaptive management and transition management. *Ecology and Society*, 16(2). Available at: www.ecologyandsociety.org/vol16/iss2/art9/.

Wan, F., Williamson, P.J. and Yin, E., 2015. Antecedents and implications of disruptive innovation: Evidence from China. *Technovation*, 39–40, pp. 94–104.

Weston, F. and Dupuy, M., 2019. Five suggestions to promote China power sector transition. Available at: www.raponline.org/blog/five-suggestions-to-china-power-sector-reform/ (Accessed 27 May 2019).

Xiao, Y., Tylecote, A. and Liu, J., 2013. Why not greater catch-up by Chinese firms? The impact of IPR, corporate governance and technology intensity on late-comer strategies. *Research Policy*, 42(3), pp. 749–764.

Ydersbond, I.M. and Korsnes, M.S., 2016. What drives investment in wind energy? A comparative study of China and the European Union. *Energy Research & Social Science*, 12, pp. 50–61.

Zhang, S., Andrews-Speed, P. and Li, S., 2018. To what extent will China's ongoing electricity market reforms assist the integration of renewable energy?. *Energy Policy*, 114, pp. 165–172.

Zhou, Y., Pan, M. and Urban, F., 2018. Comparing the international knowledge flow of China's wind and solar photovoltaic (pv) industries: Patent analysis and implications for sustainable development. *Sustainability*, 10(6), 1883.

Zhu, M., Qi, Y., Belis, D., Lu, J. and Kerremans, B., 2019. The China wind paradox: The role of state-owned enterprises in wind power investment versus wind curtailment. *Energy Policy*, 127, pp. 200–212.

Appendix

Overview of interviews

#	*Organisation/ industry segment*	*Date*	*Position*	*Origin of organisation*	*Place of interview*
1	Policy and advisory	12/09/2011	China director	Europe	Beijing
2	Wind and solar developer	15/09/2011	Advisor	China	Beijing
3	Policy and advisory	26/09/2011	Project manager, renewable energy	Europe	Beijing
4	Policy and advisory	29/09/2011	Vice secretary-general	China	Beijing
5	Government institute	30/09/2011	Deputy director general	China	Beijing
6	Turbine manufacturer	27/10/2011	Key account manager	China	Beijing
7	Turbine manufacturer	31/10/2011	Sales manager	China	Beijing
8	Policy and advisory	01/11/2011	Wind analyst	US	Beijing
9	Turbine manufacturer	03/11/2011	Marketing manager	Europe	Beijing
10	Policy and advisory	08/11/2011	Director of research and advisory	Europe	Beijing
11	Government institute	08/11/2011	Research associate on economy and energy policy	China	Beijing
12	Turbine manufacturer	14/11/2011	Technical support engineer	China	Beijing
13	Turbine manufacturer	03/07/2013	Chief engineer	China	Shanghai
14	Certification and advisory	20/08/2013	Head of renewable energy service	Europe	Beijing
15	Foundation coating	28/10/2013	CEO	China	Shanghai
16	Gear, Pitch, Yaw	30/10/2013	Sales manager	Europe	Shanghai
17	Turbine manufacturer	30/10/2013	Business developer	Europe	Shanghai

#	*Organisation/ industry segment*	*Date*	*Position*	*Origin of organisation*	*Place of interview*
18	Coating	31/10/2013	Sales manager	Europe	Shanghai
19	Certification and advisory	31/10/2013	Manager China offshore and certification engineer	Europe	Shanghai
20	Certification and advisory	01/11/2013	Manager – wind energy	Europe	Shanghai
21	Turbine manufacturer	16/11/2013	Director	China	Shanghai
22	Control systems	22/11/2013	General manager of China	Europe	Shanghai
23	Research/ University	06/12/2013	Professor and associate professor	China	Shanghai
24	Certification	09/12/2013	Certification engineer	Europe	Shanghai
25	Policy and advisory	09/12/2013	One consultant (European) and one business developer China (Chinese)	Europe	Shanghai
26	Certification	11/12/2013	Head of renewable energy service	Europe	Beijing
27	Advisory	11/12/2013	Consultant	Europe	Beijing
28	Government	11/12/2013	Director general	China	Beijing
29	Government	11/12/2013	Attaché	Europe	Beijing
30	Government	12/12/2013	Deputy director general	China	Beijing
31	Government	12/12/2013	Advisor	China	Beijing
32	Government	12/12/2013	Executive vice secretary-general	China	Beijing
33	Turbine manufacturer	13/12/2013	Offshore wind structural engineer	China	Beijing
34	Advisory	13/12/2013	Consultant	Europe	Beijing
35	Advisory	18/12/2013	Consultant	Europe	Copenhagen
36	Advisory	20/01/2014	Regional manager	Europe	Shanghai
37	Government	26/02/2014	Deputy director and assistant	China	Nanjing
38	Gearbox	26/02/2014	Account manager	China	Chang-zhou
39	Developer	11/03/2014	Vice director of preparatory department	China	Hangzhou
40	Turbine manufacturer	11/03/2014	Offshore wind technical director and control system supervisor	China	Hangzhou

continued

Appendix Continued

#	*Organisation/ industry segment*	*Date*	*Position*	*Origin of organisation*	*Place of interview*
41	Cable	17/03/2014	Sales manager	China	Nantong
42	Installation, Foundation, Port	17/03/2014	Senior business manager offshore wind	China	Qidong
43	Construction and Operation	01/04/2014	Head of engineering department	China	Rudong
44	Turbine manufacturer	01/04/2014	Control system engineer, offshore wind turbine	China	Rudong
45	Government	02/04/2014	Section leader of offshore wind planning and offshore wind construction official	China	Rudong
46	Coating	04/04/2014	Sales manager	Europe	Shanghai
47	Turbine manufacturer	04/04/2014	Technical project manager	Europe	Shanghai
48	Government	10/04/2014	Director industry park development department	China	Shanghai
49	Forecasting	14/04/2014	General manager and chief engineer China	Europe	Beijing
50	Grid	14/04/2014	Assistant director of renewable energy department	China	Beijing
51	Certification agency	15/04/2014	General manager of wind energy dept.	China	Beijing
52	Advisory	15/04/2014	General manager of China division	Europe	Beijing
53	Developer	16/04/2014	Deputy director engineer and development deputy.	China	Beijing
54	Developer	17/04/2014	Director, production and operation department	China	Zhuhai

#	Organisation/ industry segment	Date	Position	Origin of organisation	Place of interview
55	Developer	17/04/2014	Department of resources development, offshore wind	China	Guangzhou
56	University	18/04/2018	Researcher	Europe	Beijing
57	University	22/04/2018	Professor	China	Shanghai
58	University	23/04/2018	Professor	China	Shanghai
59	Government institute	28/04/2018	Advisor	China	Beijing
60	University	28/04/2018	Professor	China	Beijing
61	Policy and advisory	03/05/2018	Analyst	China	Beijing
62	Policy and advisory	03/05/2018	Analyst	Europe	Beijing
63	Government institute	05/05/2018	Senior analyst	China	Beijing
64	University	10/05/2018	Professor	Europe	Beijing
65	University	10/05/2018	Professor	China	Beijing
66	Policy and advisory	11/05/2018	Energy specialist	International	Beijing
67	Policy and advisory	11/05/2018	Vice president of strategy	Europe	Beijing
68	Policy and advisory	25/05/2018	Analyst	Europe	Beijing
69	Government institute	29/05/2018	Energy advisor	China	Beijing
70	University	31/05/2018	Professor	China	Beijing
71	PV company	11/06/2018	Manager	China	Beijing
72	University	11/06/2018	Professor	China	Beijing
73	PV company	12/06/2018	Project developer	China	Beijing
74	Research and Advisory Institute	21/06/2018	Engineer	China	Beijing
75	PV company	22/06/2018	Engineer	China	Shunyi, Beijing
76	University	17/08/2018	Professor	China	Trondheim
77	Government institute	24/04/2019	Senior advisor, energy and environment	Europe	Trondheim (Skype)

Index

Page numbers in *italics* denote figures, those in **bold** denote tables.

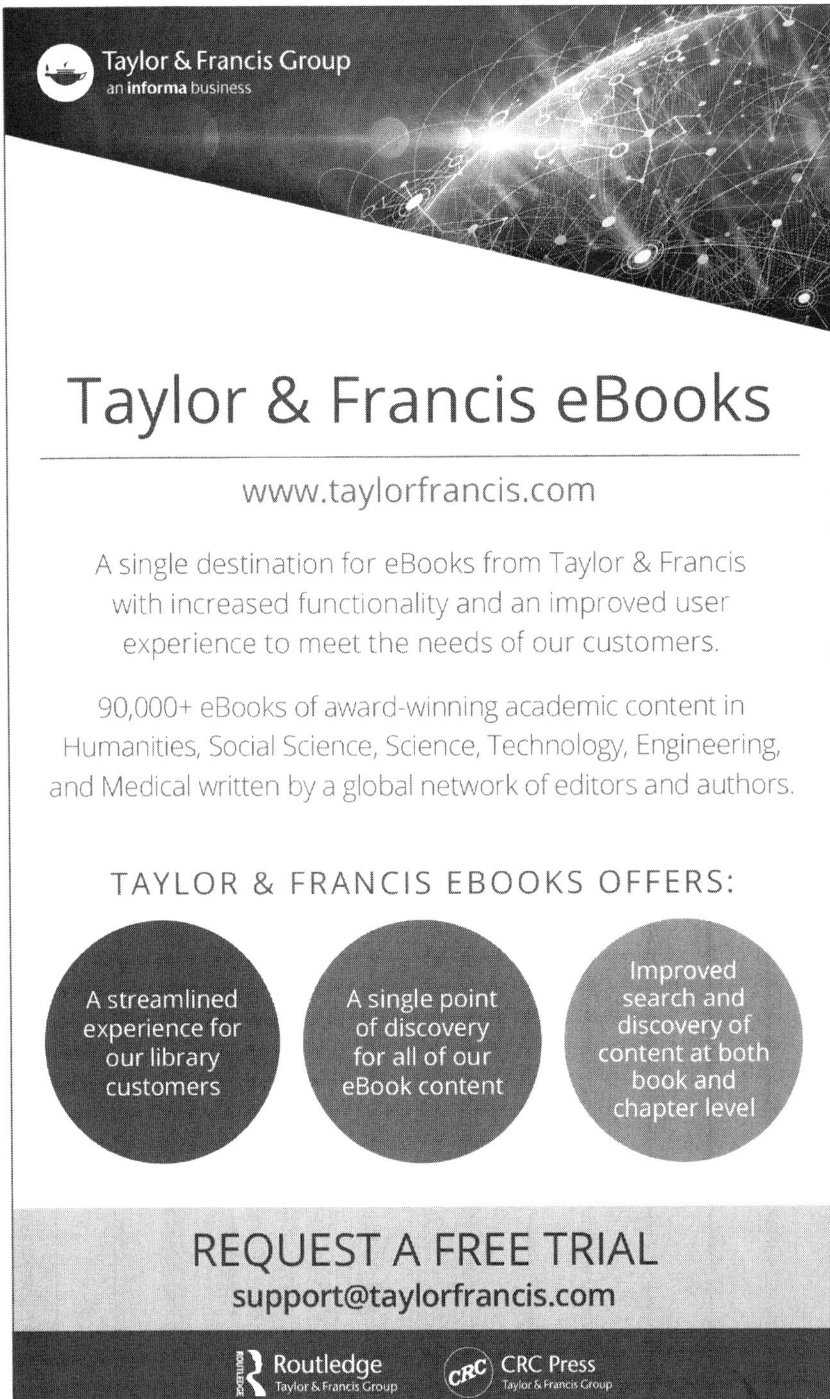
Taylor & Francis Group
an informa business
Taylor & Francis eBooks
www.taylorfrancis.com
A single destination for eBooks from Taylor & Francis with increased functionality and an improved user experience to meet the needs of our customers.
90,000+ eBooks of award-winning academic content in Humanities, Social Science, Science, Technology, Engineering, and Medical written by a global network of editors and authors.
TAYLOR & FRANCIS EBOOKS OFFERS:
A streamlined experience for our library customers
A single point of discovery for all of our eBook content
Improved search and discovery of content at both book and chapter level
REQUEST A FREE TRIAL
support@taylorfrancis.com
Routledge
Taylor & Francis Group
CRC Press
Taylor & Francis Group